Lecture Notes in Artificial Intelligence   11016

Subseries of Lecture Notes in Computer Science

LNAI Series Editors

Randy Goebel
University of Alberta, Edmonton, Canada
Yuzuru Tanaka
Hokkaido University, Sapporo, Japan
Wolfgang Wahlster
DFKI and Saarland University, Saarbrücken, Germany

LNAI Founding Series Editor

Joerg Siekmann
DFKI and Saarland University, Saarbrücken, Germany

More information about this series at http://www.springer.com/series/1244

Kenichi Yoshida · Maria Lee (Eds.)

Knowledge Management and Acquisition for Intelligent Systems

15th Pacific Rim Knowledge Acquisition Workshop, PKAW 2018
Nanjing, China, August 28–29, 2018
Proceedings

Springer

Editors
Kenichi Yoshida
University of Tsukuba
Tokyo
Japan

Maria Lee
Shih Chien University
Taipei City
Taiwan

ISSN 0302-9743   ISSN 1611-3349 (electronic)
Lecture Notes in Artificial Intelligence
ISBN 978-3-319-97288-6   ISBN 978-3-319-97289-3 (eBook)
https://doi.org/10.1007/978-3-319-97289-3

Library of Congress Control Number: 2018949383

LNCS Sublibrary: SL7 – Artificial Intelligence

This Springer imprint is published by the registered company Springer Nature Switzerland AG
The registered company address is: Gewerbestrasse 11, 6330 Cham, Switzerland

Preface

This volume contains the papers presented at the 2018 Pacific Rim Knowledge Acquisition Workshop (PKAW 2018) held in conjunction with the 15th Pacific Rim International Conference on Artificial Intelligence (PRICAI 2018), during August 28–29, 2018 in Nanjing, China.

Artificial intelligence (AI) research has evolved over the past few decades and knowledge acquisition research is one of the cores of AI research. Three international knowledge acquisition workshops have been held in the Pacific Rim, Canada, and Europe since the 1990s. Over the years, AI and knowledge acquisition have adopted many technologies and flourished. Industries aided in this way have prospered and we hope they will continue to thrive. We invited the PKAW co-founders and honorary chairs, Prof. Paul Compton and Prof. Hiroshi Motoda, to share and provide visionary talks on "PKAW: From Past to Future" and "Social Network as a Rich Source of Human Behavior" at PKAW 2018.

PKAW 2018 had a strong emphasis on incremental knowledge acquisition, machine learning, deep learning, social network analysis, big data, data mining, and agents. The proceedings contain 15 regular papers and seven short papers that were selected by the Program Committee among 51 submitted papers. All papers were peer-reviewed by three reviewers. The papers in these proceedings cover the methods and tools as well as the applications related to developing a knowledge base, health care, financial systems, and intelligent systems.

The workshop co-chairs would like to thank all those who contributed to PKAW 2018, including the PKAW Program Committee and reviewers for their support and timely review of papers, and the PRICAI Organizing Committee for handling all of the administrative and local matters. Thanks to EasyChair for streamlining the whole process of producing this volume and Springer for publishing the proceedings in the *Lecture Note in Artificial Intelligence* (LNAI) series. Particular thanks to those who submitted papers, presented, and attended the workshop. We look forward to seeing you at PKAW 2020.

August 2018

Kenichi Yoshida
Maria Lee

Organization

Program Committee

Nathalie Aussenac-Gilles	IRIT, CNRS and University of Toulouse, France
Quan Bai	Auckland University of Technology, New Zealand
Ghassan Beydoun	University of Technology, Sydney, Australia
Xiongcai Cai	The University of New South Wales, Australia
Tsung Teng Chen	National Taipei University, Taiwan
Jérôme David	Inria, France
Akihiro Inokuchi	Kwansei Gakuin University, Japan
Toshihiro Kamishima	National Institute of Advanced Industrial Science and Technology (AIST), Japan
Byeong-Ho Kang	University of Tasmania, Australia
Mihye Kim	Catholic University of Daegu, South Korea
Yang Sok Kim	Keimyung University, South Korea
Alfred Krzywicki	The University of New South Wales, Australia
Maria Lee	Shin Chien University, Taiwan
Kyongho Min	The University of New South Wales, Australia
Toshiro Minami	Kyushu Institute of Information Sciences and Kyushu University Library, Japan
Luke Mirowski	University of Tasmania, Australia
Tsuyoshi Murata	Tokyo Institute of Technology, Japan
Kouzou Ohara	Aoyama Gakuin University, Japan
Hayato Ohwada	Tokyo University of Science, Japan
Tomonobu Ozaki	Nihon University, Japan
Hye-Young Paik	The University of New South Wales, Australia
Ulrich Reimer	University of Applied Sciences St. Gallen, Switzerland
Deborah Richards	Macquarie University, Australia
Kazumi Saito	University of Shizuoka, Japan
Derek Sleeman	University of Aberdeen, UK
Vojtěch Svátek	University of Economics, Prague, Czech Republic
Takao Terano	National Institute of Advanced Institute of Science and Technology (AIST), Japan
Hiroshi Uehara	Akita Prefectural University, Japan
Shuxiang Xu	University of Tasmania, Australia
Takahira Yamaguchi	Keio University, Japan
Kenichi Yoshida	University of Tsukuba, Japan
Tetsuya Yoshida	Nara Women's University, Japan

Contents

Building a Commonsense Knowledge Base for a Collaborative Storytelling Agent

Dionne Tiffany Ong, Christine Rachel De Jesus,
Luisa Katherine Gilig, Junlyn Bryan Alburo, and Ethel Ong$^{(\boxtimes)}$

De La Salle University, Manila, Philippines
{dionne_ong, christine_dejesus, ethel.ong}@dlsu.edu.ph

Abstract. Storytelling is a common activity that people engage in to share and exchange information about their everyday life events. Children in particular find storytelling entertaining as they learn about their world and even share their own stories. Virtual agents are gaining popularity as conversational agents who can engage with their human users in a dialogue to answer queries and to find the necessary support in the performance of some tasks in a particular application domain. For virtual agents to be able to share stories with their human users, specially the children, they need to be provided with a body of knowledge to use in their dialogue. In this paper, we describe our approach in building an ontology to provide a knowledge base of commonsense concepts and their relations to a virtual agent. This is then used by a virtual agent to process user input as a form of story text, and to generate appropriate responses in order to encourage the child to share his/her story.

Keywords: Commonsense ontology · Virtual agent · Storytelling
Dialogue

1 Introduction

Stories are part of our everyday lives. We use stories to communicate our experiences and to empathize with others. As a universal habit [1], storytelling sessions are used in educating positive attitudes such as respect and openness [2]. When shared orally, stories are also effective in enhancing different communication skills such as articulation, enunciation, and vocabulary as reported in the study of [3].

Stories are usually shared with an audience who spins their own understanding onto a story, whether out loud or not, thus becoming part of the storytelling process [4]. This is evident even with young children, who may be silent at the onset of the activity before turning into storytellers themselves by sharing their own personal recollections and interpretations with the main storyteller [5]. When encountered by situations such as writer's block, which is a common occurrence in the production of any creative artifact [6], storytellers often find themselves soliciting ideas from their audience. As the questions, interpretations and feedback are raised, the story evolves and becomes complete through this exchange of viewpoints among the collaborators.

In the field of computing, researchers in human-computer collaboration are exploring various approaches in modelling collaboration and designing collaborative

© Springer Nature Switzerland AG 2018
K. Yoshida and M. Lee (Eds.): PKAW 2018, LNAI 11016, pp. 1–15, 2018.
https://doi.org/10.1007/978-3-319-97289-3_1

systems. A survey conducted by Kybartas and Bidarra [7] reported the use of "mixed initiative" as a collaborative strategy to engage computers actively in the development of various narrative forms. While their study examined the varying degrees for automating certain steps in the story generation process, they also emphasized that "the range of stories which may be generated (by the computer) is still largely dependent on the information" it has at its disposal, or supplied to it by a human author.

Storytelling between a human user and a computer requires more than collaborative spaces where the user is immersed in the virtual world or where users are given tools for the sharing and exchange of artifacts. Full collaboration requires spaces that allow tasks such as the generation of ideas, giving feedback and suggestions, and even co-authoring, to flourish.

For many collaborative systems, virtual agents are used as the conduit for human-computer interaction. Virtual agents are intelligent characters designed to collaborate with human users to help them think better during the performance of specific tasks. They utilize various means of communication, from written to verbal to non-verbal, in order to "engage (their human users) in dialogues, and negotiate and coordinate the transfer of information" [8].

One such agent is the chatbot. A chatbot mainly uses text understanding and generation techniques to communicate with its users. Chatbots have been designed to assist users in different application domains, specifically those that employ a question-and-answer type of interaction to allow chatbots to cater to the needs of users from booking air tickets and hotel accommodations, to facilitating activities in learning and game-based environments. In educational applications, chatbots fulfill a variety of roles, including that of a critic who looks over the shoulder of the users as they perform their tasks, and offers appropriate advices as the need arises [9]; a tutor who encourages students to perform the required learning task and clarifies learner misconceptions [10]; and a virtual peer or playmate who engages students in learning-based conversations to promote a two-way transmission of knowledge [11]. While these learning sessions can be partially thought of as collaborative, none of them actually allow the virtual agent to become a co-creator. Rather, chatbots currently serve the role of helpers or assistants to humans, performing commands such as playing music, controlling room lights, and searching the web to find answers to a user query.

To be an effective storytelling partner, virtual agents must be endowed with a collection of knowledge that include commonsense concepts about our world and narrative knowledge on story structures and storytelling. Commonsense knowledge can enable computers to "reason in useful ways about ordinary human life" [12] by possessing knowledge about our everyday activities and places where we perform them, objects found in our world, causal chain of events, and our inter-relationships.

In this paper, we describe our approach in populating a commonsense knowledge resource with everyday concepts and their relations that are familiar to young children. We explain the challenges we encountered in sourcing knowledge from existing resources and how we address these challenges. We also discuss briefly how the resulting knowledge base is used by our virtual agent to engage children in dialogues meant to encourage them to share their stories.

2 Related Works

Ein-Dor [13] presented different types of commonsense knowledge. This includes generally known facts about the world, knowledge on human behaviors and their usual outcomes, relating causes and effects (of actions and events), and being able to solve what people consider "everyday problems".

For a virtual agent whose goal is to encourage a child to share his/her stories, commonsense knowledge is needed to generate relevant responses to given inputs. This means that general knowledge must be stored in the form of assertions that computers can use and reason with. By making these assertions accessible, the agent is able to formulate responses that are aligned with the input given by the user.

Various research works have looked into the development of large-scale commonsense knowledge resources to give machines the ability to reason about everyday human life. SUMO, or the Suggested Upper Merged Ontology [14], is an open source formal and public ontology owned by IEEE that captures the semantics of commonsense world knowledge as concepts represented as axioms in first-order logic form. The work of [15] has investigated SUMO as a formal representation of storytelling knowledge and story plans, while its inference engine Sigma has been used to query events that may be added into the story being generated.

Cyc [16] is a language-independent and extensible knowledge base and commonsense reasoning engine that contains about 24.5 million proprietary and handcrafted assertions. FrameNet [17] is a free lexical database developed at Berkeley with annotated examples showing word meaning and usage, mostly used for semantic role labelling and information extraction. WordNet [18] is another free lexical database developed at Princeton and contains nouns, verbs, adjectives and adverbs representing concepts that are organized into synonyms or synsets. These systems, however, lack the informal relations between everyday concepts that may abound during storytelling with children.

In 2004, researchers at the MIT Media Lab began building several large-scale commonsense knowledge bases that contain descriptions of human life, including the actions people can take and their corresponding effect, and the different objects we can find in our world and their attributes and usage. These systems include ConceptNet, LifeNet and StoryNet [12].

In this paper, we only look at ConceptNet [19] which contains assertions that can enable machines to reason about event occurrences and object relationships. This type of reasoning has been explored in the MakeBelieve [20] story generation system which performs logical reasoning on the cause and effect relations of a chain of events found in ConceptNet. Commonsense reasoning has also been used in designing intelligent interface agents, as described in [21]. In these applications, the agent is tasked to provide help, assistance and suggestions by inferencing the commonsense knowledge base to find answers to user queries.

3   Extracting Knowledge from ConceptNet

ConceptNet [22] represents commonsense knowledge in a semantic network consisting of word relations that include both the lexical definitions as well as other concepts or words associated with said word. Nodes in the network contain concepts, while edges connecting two nodes denote their semantic relations. This representation of knowledge as a binary assertion can be written as

```
[concept1 semantic-relation concept2]
```

Concepts *concept1* and *concept2* are simple words or phrases that refer to real-world entities or ideas, while the semantic relation defines how these concepts are related by their lexical definitions and through their common knowledge. Example assertions include information about everyday objects, such as *['cake' IsA 'food'], ['cake', HasProperty 'sweet']* and *['spoon' UsedFor 'eating']*; event relations such as *['eat' HasSubevent 'chew']* and *['eat' AtLocation 'restaurant']*; and motivations and desires, such as *['person' Desires 'happy']* and *['eat' MotivatedByGoal 'survive']*. A semantic relation can usually be translated into a common English sentence pattern, as shown in Fig. 1, and use in template-based text generation in various application domains.

Relation	Sentence pattern	Relation	Sentence pattern
IsA	*NP* is a kind of *NP*.	LocatedNear	You are likely to find *NP* near *NP*.
UsedFor	*NP* is used for *VP*.	DefinedAs	*NP* is defined as *NP*.
HasA	*NP* has *NP*.	SymbolOf	*NP* represents *NP*.
CapableOf	*NP* can *VP*.	ReceivesAction	*NP* can be *VP*.
Desires	*NP* wants to *VP*.	HasPrerequisite	*NP*|*VP* requires *NP*|*VP*.
CreatedBy	You make *NP* by *VP*.	MotivatedByGoal	You would *VP* because you want *VP*.
PartOf	*NP* is part of *NP*.	CausesDesire	*NP* would make you want to *VP*.
Causes	The effect of *VP* is *NP*|*VP*.	MadeOf	*NP* is made of *NP*.
HasFirstSubevent	The first thing you do when you *VP* is *NP*|*VP*.	HasSubevent	One of the things you do when you *VP* is *NP*|*VP*.
AtLocation	Somewhere *NP* can be is *NP*.	HasLastSubevent	The last thing you do when you *VP* is *NP*|*VP*.
HasProperty	*NP* is *AP*.		

Fig. 1. Examples of ConceptNet relations and corresponding sentence patterns [22].

ConceptNet's current 28 million assertions come from many sources, including those crowdsourced through the Commonsense Computing Initiative, DBPedia, Wiktionary, Games with a Purpose, WordNet, OpenCyc and Umbel [19]. Because of this, there exist advanced and even negative concepts that may not be relevant nor appropriate when interacting with young children, specifically those between 7 to 10 years old. This constraint led us to build a separate ontology of commonsense knowledge that adapts the semantic relations of ConceptNet, but without its inappropriate and sometimes harmful content.

To extract semantic knowledge from ConceptNet, its Web API must be used. This Web API returns a JSON string containing information on a specified concept or concept-relation pair. A ConceptNet API was used to convert this JSON data into a more easily parsable string with the *query()* function. This function takes in a word that serves as the general concept being extracted, a specific relation for that word, and a node position which specifies if the unknown concept was to be at the first (start) or the second concept. The function then returns a list of arrays containing the extracted concepts. For example, given *query("bee", "isA", "start")*, ConceptNet returns the list *[('cuckoo-bumblebee', 'bee'), ('drone', 'bee'), ('husking bee', 'bee'), ...].*

The *query()* function can optionally take in a weight threshold to limit the assertions to be retrieved. This weight value determines the believability of the knowledge found in ConceptNet; the higher the value, the more reliable the source is. We currently set the threshold at 2. Similar to the findings reported in [23], empirical tests showed that weights lower than 2 will yield a number of assertions that do not contain useful information, while setting a value higher than 2 will remove most of the relevant concepts from the result.

To determine which assertions to extract from ConceptNet, we created a corpus of words containing nouns, verbs, adjectives and adverbs derived from five (5) Aesop fables. Aesop fables are commonly used in young children's education and are considered a core part of children's literature [24]. The fables are *"The Princess and the Pea"*, *"A Cartload of Almonds"*, *"The Steadfast Tin Soldier"*, *"Advising a Fool"*, and *"The Tinderbox"*. No specific criteria were used in identifying these fables; the primary consideration is to build the seed corpus of words that can be fed to ConceptNet to find relevant knowledge. Furthermore, using these fables ensured that the ontology of our virtual agent would only contain concepts that have been deemed as child-friendly. It also allowed the knowledge base to include more fairy-tale like elements which children may find more interesting in their stories. Table 1 lists the sample seed words sourced from each of the fables and the corresponding assertions extracted from ConceptNet. The total number of assertions extracted is 22,285.

Table 1. Seed concepts from five (5) Aesop fables and their corresponding assertions.

Story	Sample seed	Assertions
The Princess and the Pea	prince	[prince IsA ruler] [prince AtLocation castle] [prince AtLocation England] [prince AtLocation *fairy tale]*
	marry	[priest UsedFor marry] [*start family* HasPrerequisite marry], [minister UsedFor marry] [*wedding ring* UsedFor marry]
	princess	[*princess royal* IsA princess] [*girl can fancy that she* IsA princess] [princess IsA King's daughter] [princess IsA royalty]

(continued)

Table 1. (*continued*)

Story	Sample seed	Assertions
A Cartload of Almonds	squirrel	[squirrel IsA fur] [squirrel IsA animal] [squirrel AtLocation tree] [squirrel AtLocation forest]
	forest	[rain AtLocation forest] [stick AtLocation forest] [branch AtLocation forest] [jungle IsA forest]
	leap	[leap IsA jump] [caper IsA leap] [leap IsA distance] [pounce IsA leap]
The Steadfast Tin Soldier	soldier	[soldier IsA warrior] [soldier PartOf army] [soldier AtLocation battle] [soldier AtLocation battlefield]
	lady	[lady AtLocation church] [lady IsA female] [lady IsA woman]
	dancer	[dancer AtLocation ballet] [dancer AtLocation carnival] [dancer CapableOf dance] [dancer CapableOf dance on stage]
Advising a Fool	jungle	[marmot AtLocation jungle] [koala AtLocation jungle] [vine AtLocation jungle] [beaver AtLocation jungle]
	birds	[birds AtLocation park] [birds HasProperty light] [birds CapableOf eat] [birds CapableOf grow]
	happy	[giving a gift Causes happy] [some people HasProperty happy] [writing a poem Causes happy] [Couples HasProperty happy]
The Tinderbox	dog	[dog AtLocation park] [dog HasProperty black] [dog HasProperty brown] [dog HasProperty gray]
	love	[love AtLocation family] [child Desires love] [heart UsedFor love] [love CapableOf hurt]
	magic	[a person Desires magic] [sorcery IsA magic] [white magic IsA magic] [magic UsedFor entertainment]

4 Expanding the Commonsense Ontology

Unfortunately, simply using the five (5) fables meant that many basic concepts were not included. For example, neither the concept "*animal*" nor "*sister*" was in the knowledge base since they did not appear in the particular fables. To supplement the seed concepts derived from the fables, a basic vocabulary needs to be added to the commonsense ontology. As a basis for what would be considered 'basic' vocabulary in the English language, a list that served as a guide for people who are learning English as a second language was used. This list included words related to days, months, seasons, time, colors, places, families, jobs and other basic concepts [25]. 778 words were derived from this vocabulary list, though some words are already found in the five fables. When fed to ConceptNet following the extraction method outlined before in 2.1, 18,602 additional assertions were generated.

The next challenge concerns verbs. Stories are comprised of sequences of events that characters perform to effect some changes or as a response to some other events, as well as naturally occurring phenomenon in the story world environment. Events are determined by verbs; however, the basic vocabulary list consisted mostly of nouns, not verbs. To expand the verbs, we derive a list of 331 most common verbs found in the Macmillan English Dictionary [26]. From this list, 458 additional assertions were generated. Table 2 lists the sample nouns and verbs sourced from the English dictionaries and the corresponding assertions derived from ConceptNet.

Table 2. Seed concepts from the basic vocabulary lists and their corresponding assertions.

Type	Seed word	Assertions
Nouns	hotel	[a restaurant LocatedAt hotel] [hotel UsedFor sleeping] [resort IsA hotel]
	dessert	[Ice cream IsA dessert] [a cake UsedFor dessert] [ambrosia IsA dessert]
	dad	[dad CapableOf cook well] [dad IsA father] [dad IsA a person]
Verbs	eat	[buying a hamburger Causes eat] [animal Desires eat] [eat HasPrerequisite find food]
	learn	[learn HasPrerequisite Study] [movie UsedFor learn] [a child CapableOf learn]
	think	[answer questions HasPrerequisite think] [a head UsedFor think] [Humans CapableOf think]

In children's stories, animals and objects can also be portrayed as story characters. This poses a dilemma in determining when a given noun is to be treated as an object or as a character. Using our premise that a character is any entity capable of performing some actions in the virtual story world, the previous list of basic verbs was revised to only include non-linking and non-sensory verbs. Linking verbs (i.e., *is, become, appear*) are verbs that connect a subject to information on the subject that may or may not be sentient, such as in the sentence *"The paper became wet."* Similarly, sensory verbs (i.e., *look, smell, feel, sound, taste* and *seem*) are also used to describe a subject using one of our five senses that is not necessarily sentient. Removing these verbs from the list will prevent the virtual agent from indicating that a given noun is a sentient entity, such as in the sentence *"The flowers smell fresh."*. The remaining verbs were then fed to the ontology as an assertion of the form *[character CapableOf <verb>]*, indicating that the indicated *character* can perform the corresponding *<verb>*. This works for sentences such as *"The pauper ate lunch with the prince."*, where the presence of the assertion *[pauper CapableOf eat]* informs the agent that *pauper* is a character instead of an object because it can perform the action *eat*.

261 assertions denoting character capabilities were manually inserted into the knowledge base, bringing the whole ontology to a total of 41,606 assertions.

5   Filtering the Concepts

The commonsense ontology is meant for use by a virtual agent during its conversation with children. This necessitates the need to ensure that the concepts it uses are relevant and appropriate for the target audience. In the context of this research, relevant concepts refer to those that children are familiar with as they encounter these in their everyday life. Inappropriate words, on the other hand, refer to concepts that are sexual or violent in nature. Three approaches were utilized to identify and to filter these inappropriate concepts from being added to the ontology.

The first approach deals with ambiguous words. Some basic vocabulary words are found to have multiple meanings; *"run"* can refer to the concept of movement or as a noun referring to an instance. The multiple meanings mean a word may have different associated concepts and assertions, which can range from being child friendly to outright inappropriate, as shown in Table 3.

Table 3. Assertions with ambiguous words.

Basic word	Appropriate assertions	Inappropriate assertions
fun	[playing Causes fun] [fun IsA activity]	[sex HasProperty fun] [fun AtLocation a whorehouse]
boot	[boot IsA shoe]	[boot IsA torture]
breakfast	[breakfast IsA meal]	[kill Requires eat breakfast]

A second approach in filtering ConceptNet's assertions is by limiting the types of semantic relations to be included. In its current state, ConceptNet has 29 predefined relations excluding the ones it has imported from DBPedia. These relations include general ones like *"RelatedTo"*, or irrelevant ones (in as far as storytelling with children is concerned) like *"ExternalURL"*. To determine which relations to be retained, each semantic relation is corresponded to a story element that can be used to generate story text. The resulting list contains nine (9) assertions as shown in Table 4.

Another filtering approach to address different issues encountered during the extraction process is with the use of a blacklist. The first category of words added to the blacklist are those known to denote negative meanings, such as *"war"* and *"kill"*. However, combined with a string matching function, the filtering algorithm inadvertently removed words that contain substrings of the negative short words, such as *"award"* and *"backward"*, from the commonsense ontology. To address this, strings that would be excluded no matter the surrounding text, such as *"cocaine"*, as well as those strings that would only be excluded if they exist as a separate word, such as *"war"*, are included as separate entries in the blacklist.

Table 4. Semantic relations and their corresponding story elements.

Relation	Story Element/s
IsA	Object Attribute/Character Attribute
HasProperty	Object Attribute/Character Attribute
PartOf	Object/Character
AtLocation	Object/Character/Setting
HasPrerequisite	Event
CreatedBy	Event
Causes	Event/Relationship
Desires	Character Desire
CapableOf	Character Ability

User-generated data, such as the crowdsourced knowledge of ConceptNet, tend to be noisy. It is therefore not unexpected to find assertions that only contained articles, e.g., *["a horse" AtLocation "a"]*. The presence of these article concepts are unnecessary and only clutter the knowledge base, thus, they are added to the blacklist as well.

Finally, certain semantic relations, specifically, *AtLocation*, *CapableOf* and *UsedFor*, have been noticed to contain some inappropriate concepts, as exemplified in Table 5. Since ConceptNet does not allow for queries to operate on semantic relations

Table 5. Relation specific concepts that are added to the blacklist.

Relation	Concepts
AtLocation	[a knife AtLocation your ex husbands back]
	[a knife AtLocation a stabbing victim]
CapableOf	[someone CapableOf murder]
	[person CapableOf commit crime]
UsedFor	[a knife UsedFor killing]
	[a sword UsedFor killing]

alone, the knowledge base was manually scoured for any inappropriate words. Unfortunately, none of the approaches we employed guarantee an exhaustive blacklist of inappropriate concepts.

6 Discussion

Preliminary testing was conducted with five children who are between 7 to 10 years old. The children use English as their second language and do not manifest any speech impediments. These requirements are necessary to reduce problems related to processing speech input in Google Home, the platform used for delivering our virtual agent. Informed consent forms were administered to the guardians prior to testing. Furthermore, the children and their guardians underwent a briefing session wherein the objectives of the study and the test methodology were described. The guardians were allowed to accompany the children during the actual one-on-one interaction with the virtual storytelling agent. A post-interview was also conducted to solicit feedback regarding their experience.

Test results showed that a virtual agent needs to provide more than a simple acknowledgement, such as prompting for details and contributing its share to the story, when engaging in storytelling with children. Initially, the virtual agent was designed to respond only with common acknowledgements like *"I see."*, *"That's right."*, and *"Tell me more."*. Since responses were limited, there were occurrences wherein the same acknowledgement was used more than once in a storytelling session. With the realization that the agent's responses had no context, the children disregarded or ignored the virtual agent's utterances. Some children even kept on talking over the agent to continue their story.

During storytelling, conversations happen naturally as the narrators tell their stories and the listeners offer a variety of responses, including but not limited to, acknowledging the story text, recounting a similar story, agreeing with the narrator's story, or resisting the narrator's views [27]. Conversations are the means by which ideas are exchanged among the participants. A dialogue is a kind of conversation wherein the participants work together towards a goal; in this case, the goal is to construct a story.

A virtual agent can exhibit the same behavior as a human storyteller by utilizing the commonsense ontology during a dialogue. There are certain dialogue strategies that help promote a collaborative environment for children. These strategies were identified in [28] through analyzing the patterns employed by tutors during tutoring sessions. *Pumps* can be used when the user's initial story text is found to be lacking. They work by encouraging the user to express additional information or details to expand the story element, such as the character's attributes, or the consequential event. *Hints* can be used when the user is stuck with his/her current story. They work by mentioning a fact or anything related to the previous story text shared by the user to help trigger ideas.

6.1 Processing User Input

For a virtual agent to use the dialogue strategies and get the child to continue telling the story, it must be able to provide relevant feedback that is not limited to what the child has already said. The agent can make use of the ontology which now contains a variety of concepts, from basic concepts retrieved from the vocabulary lists to more domain-specific ones sourced from the fables to process user inputs.

When processing an input text, text understanding methods that depend mostly on deciphering the grammatical structure of a sentence are not sufficient to correctly identify the semantic role of words found in the sentence. A word's associated concept needs to be taken into account to help the agent understand the role of the word or phrase in the context of the story. The ontology can be used to supplement the agent's text understanding module.

One application of this is in determining if a simple object mentioned in the story text is a sentient character. Magical themes can often be found in children's stories, for example, in the fairy tale "*Beauty and the Beast*", the supporting cast is comprised mostly of enchanted objects, e.g., the candelabra *Lumiere*, the clock *Cogsworth*, the teapot *Mrs. Potts* and the tea cup *Chip*. Because of these themes, we cannot discount typically non-living objects from being posed as story characters. One of the ways to distinguish between an object and a sentient character is to look at the verb used in a sentence which can determine the action that an object is capable of performing. If an object is capable of doing an action that only sentient beings can do, they will be considered a character rather than an object.

Consider the input text "*The statue was stone grey and it walked off the pedestal.*" POS taggers that analyze the sentence structure will identify the linking verb *was* and *walked* as verbs. But for purposes of storytelling, the concept of a verb denoting an *action* rather than *describing* its subject needs to be differentiated. The virtual agent needs to know that *statue* is a possible story character because it performed an action *walk*. This knowledge should come from the commonsense ontology, where the agent queries for assertions of the form *[character CapableOf character-ability]* to find that *walk* is a character ability.

Another application of the ontology is in determining if a given noun denotes the location. Locations are often indicated an an object of preposition, i.e., "*in the library*" and "*to the park*". However, there are different types of prepositional objects, including other objects ("*in the bag*") and time ("*in the afternoon*"). To help the agent in determining if a given concept is a location, it queries the commonsense ontology for assertions of the form *[<noun> IsA place]* or *[<noun> IsA location]*. Consider the two input sentences, "*She went to eat.*" and "*She went to school.*". Both have the preposition *to* with the prepositional objects *eat* and *school*, respectively. When the ontology is queried, only the assertion *[school IsA place]* is found and "*school*" is tagged as a location accordingly.

6.2   Generating Responses

The commonsense ontology is also used by the virtual agent in order to contribute relevant information when constructing a response to user inputs. By expounding on a child's story text, related concepts and objects can be added into the story world to help the child continue with his/her story especially in situations when he/she is suffering from a writer's block and needs help to add new details to the story.

Using the sample assertions in Tables 1 and 2, and the relation-to-story element correspondence in Table 4, the virtual agent should be able to assist the child through different dialogue moves called responses that utilize hints and pumps. Table 6 lists some example input text and possible agent response.

Table 6. Example agent responses to user input text.

User input	Agent response	Response type
#1 The prince is tired.	The prince went to the castle.	hint
#2 The princess is a royalty.	The princess loves the prince.	hint
#3 The chef cooked a burger.	What happened to the burger?	pump
#4 The chef wants an apple. He bought one apple.	What will the chef do next?	pump

Example #1 in Table 6 illustrates the virtual agent's ability to process that story characters can have attributes by using the assertion form *[character HasProperty character-attribute]*, and subsequently generates a hint by using another assertion *[character AtLocation setting]*. It uses the knowledge *[prince AtLocation castle]* from the ontology to determine a candidate setting/location for the *prince*.

In the second example, the character attribute from the semantic relation *[character IsA character-attribute]* is determined from the assertion *[princess IsA royalty]* given in Table 2, which is then linked to the character's capabilities relation *[character CapableOf character-ability]*. Querying the ontology leads to the assertion *[character CapableOf love]* that is then used by the agent to generate a hint.

Examples #3 and #4 typifies the use of non-fairy tale concepts and pumps as a response strategy. The input text in #3, *"The chef cooked a burger."*, is processed using the assertion form *[object CreatedBy character-action]*, which supports the story event element. The agent uses a pump to ask for more details regarding the event *"cook"* using the assertion form *[character-action Causes character-action]*.

In example #4, *"The chef wants an apple. He bought one apple."*, the virtual agent uses the semantic relation *[character Desires object]* to process the input. The second sentence is processed with the assertion form *[action HasPrerequisite action]*, causing the agent to generate a pump to ask the user to expound on the event *"buy"*.

7 Conclusion and Further Work

Oral storytelling encourages children to express themselves through their stories by engaging them in conversations regarding the setting, characters and events taking place in the narrative. This builds their creativity while also developing their literacy and communication skills. Virtual agents have been employed in various domains to support human users in various tasks, from online retail applications to healthcare diagnosis. In learning environments, virtual agents can facilitate learning activities, clarify misconceptions and even serve as learning companions to children.

As chatbots and conversational agents become more involved in supporting us perform our everyday tasks, we explore in this paper their potential as storytelling companions of children. Doing so may enable them to exhibit a form of machine intelligence where they embody certain social skills and everyday knowledge to participate in human activities in more useful ways. But first, they must be given a body of commonsense knowledge about our world that they can use when engaging children in storytelling dialogue.

In this paper, we rationalized the need for building our own ontology of commonsense concepts and relations familiar to children. We also described challenges we encountered when sourcing the knowledge from existing resources, specifically ConceptNet, to populate our knowledge base with relevant concepts appropriate for children. We want our ontology to contain commonsense knowledge that model realities found in our world while at the same time embody some elements of fairy tale fantasy commonly exhibited by children's stories.

In its current iteration, our ontology of commonsense knowledge contains 41,606 general-purpose assertions. We have briefly shown how this ontology can be used by an agent to supplement its knowledge when processing user input and generating responses. Responses are generated in the form of pumps that ask for additional details to expand the story text, as well as hints to trigger story ideas. As shown in our discussions, the responses are currently limited to single sentences. We are working on expanding the usage of hints by posing these as suggested story text, i.e., "*What about saying – the princess went to the palace?*", thus promoting collaboration while emphasizing ownership of the user to the shared story. Other dialogue strategies, such as providing feedback on the input text, and allowing users to ask questions, are also being considered to increase the variances in the agent's responses.

Meehan [29] pointed out that it is the lack of knowledge that lead computers to generate "wrong" stories. Virtual agents with a limited knowledge base may also suffer from "writer's block", which can lead children to find them as ineffective storytelling companions. There is no explicit means by which we can measure how extensive the knowledge base should be in order to cover all possible story themes that children may want to share with the virtual agent.

Aside from conducting further testing with the target audience to gauge the agent's performance as a storytelling companion, future work can consider a facility that will enable the virtual agent to determine a gap in its knowledge when it cannot process a new concept found in the user's input. Once this is detected, the agent should acquire and store this concept as a new assertion. It then utilizes current dialogue strategies, such as pumps, to get more details from the user in order to expand its knowledge about the new concept. Because crowdsourced knowledge need to be verified, the agent then uses hints to increase its "vote of confidence" on the correctness or truthfulness of the new knowledge.

Finally, various studies [30] have identified different types of knowledge that are needed to support a range of collaborative input processing and text generation tasks within the storytelling context, and domain-based commonsense knowledge is just one of them. Our ongoing work includes maintaining a story world model to track story elements, namely, characters, objects and setting (time and place), that have been identified and extracted from user inputs. The model is updated dynamically as new inputs are received from the user, such as story text that details the attributes of story characters and objects. An event chain is also used to track the sequence of events that the entities in the story go through, as narrated by the user. The storytelling agent will subsequently use the story world model and the event chain to generate a story from its conversation with the user.

References

1. Meek, M.: On Being Literate. Penguin Random House, London (2011)
2. Abma, T.: Learning by telling: Storytelling workshops as an organizational learning intervention. Manag. Learn. **34**(2), 221–240 (2003)
3. Mokhtar, N.H., Halim, M.F.A., Kamarulzaman, S.Z.S.: The effectiveness of storytelling in enhancing communicative skills. Procedia-Soc. Behav. Sci. **18**, 163–169 (2011)
4. Kosara, R., Mackinlay, J.: Storytelling: The next step for visualization. Computer **46**(5), 44–50 (2013)
5. Lawrence, D., Thomas, J.: Social dynamics of storytelling: Implications for story-base design. In: Proceedings of the AAAI 1999 Fall Symposium on Narrative Intelligence, North Falmouth, MA (1999)
6. Rose, M.: Writer's Block: The Cognitive Dimension. Studies in Writing and Rhetoric. Southern Illinois University Press, Carbondale (2009)
7. Kybartas, B., Bidarra, R.: A survey on story generation techniques for authoring computational narratives. IEEE Trans. Comput. Intell. AI Game **9**(3), 239–253 (2017)
8. Coen, M.H.: SodaBot: A Software Agent Construction System. Massachusetts Institute of Technology AI Lab (1995)
9. Terveen, L.G.: An overview of human-computer collaboration. Knowl. Based Syst. **8**(2–3), 67–81 (1995)
10. Graesser, A.C., Olney, A., Haynes, B.C., Chipman, P.: AutoTutor: an intelligent tutoring system with mixed-initiative dialogue. IEEE Trans. Educ. **48**(4), 612–618 (2005)
11. Cassell, J., Tartaro, A., Rankin, Y., Oza, V., Tse, C.: Virtual peers for literacy learning. Educ. Technol. **47**(1), 39–43 (2005). Special issue on Pedagogical Agents
12. Singh, P., Barry, B., Liu, H.: Teaching machines about everyday life. BT Technol. J. **22**(4), 227–240 (2004)
13. Ein-Dor, P.: Commonsense knowledge representation I. In: Encyclopedia of Artificial Intelligence, pp. 327–333. IGI Global (2009)
14. Suggested Upper Merged Ontology (SUMO). http://www.adampease.org/OP/. Accessed 11 Apr 2018
15. Cua, J., Ong, E., Manurung, R., Pease, A.: Representing story plans in SUMO. In: NAACL Human Language Technology 2010 Second Workshop on Computational Approaches to Linguistic Creativity, pp. 40–48. ACL, Stroudsburg (2010)
16. Cyc: Logical reasoning with the world's largest knowledge base. http://www.cyc.com. Accessed 11 Apr 2018
17. FrameNet. https://framenet.icsi.berkeley.edu/fndrupal/. Accessed 11 Apr 2018
18. WordNet: A lexical database for English. https://wordnet.princeton.edu. Accessed 11 Apr 2018
19. ConceptNet: An open, multilingual knowledge graph. http://conceptnet.io. Accessed 11 Apr 2018

20. Liu, H., Singh, P.: Makebelieve: using commonsense knowledge to generate stories. In: Proceedings of the 18th National Conference on Artificial Intelligence, Edmonton, pp. 957–958. AAAI (2002)
21. Lieberman, H., Liu, H., Singh, P., Barry, B.: Beating some common sense into interactive applications. In: AI Magazine. AAAI Press (2004)
22. Speer, R., Havasi, C.: Representing general relational knowledge in ConceptNet 5. In: Choukri, K., Declerck, T., Dogan, M.U., Maegaard, B., Mariani, J., Odijk, J., Piperidis, S. (eds.) Proceedings of the 8th International Conference on Language Resources and Evaluation (LREC 2012), Istanbul, Turkey (2012)
23. Yu, S., Ong, E.: Using common-sense knowledge in generating stories. In: Anthony, P., Ishizuka, M., Lukose, D. (eds.) PRICAI 2012. LNCS (LNAI), vol. 7458, pp. 838–843. Springer, Heidelberg (2012). https://doi.org/10.1007/978-3-642-32695-0_82
24. Lerer, S.: Children's literature: A reader's history, from Aesop to Harry Potter. University of Chicago Press, Chicago (2009)
25. Basic ESL Vocabulary. http://www.esldesk.com/vocabulary/basic. Accessed 10 Apr 2018
26. Rundell, M. (ed.): Macmillan Essential Dictionary: For Learners of English. Macmillan, Oxford (2003)
27. Mandelbaum, J.: Storytelling in conversation. In: Sidnell, J., Stivers, T. (eds.) The Handbook of Conversation Analysis, pp. 492–507. Blackwell Publishing Ltd. (2013)
28. Graesser, A.C., Wiemer-Hastings, K., Wiemer-Hastings, P., Kreuz, R., Group, T.R.: Autotutor: a simulation of a human tutor. Cogn. Syst. Res. 1(1), 35–51 (1999)
29. Meehan, J.: Tale-Spin - An interactive program that writes stories. In: Proceedings of the 5th International Joint Conference on Artificial Intelligence, pp. 91–98. Morgan Kaufmann Publishers Inc, CA (1977)
30. Oinonen, K., Theune, M., Nijholt, A., Uijlings, J.: Designing a story database for use in automatic story generation. In: Harper, R., Rauterberg, M., Combetto, M. (eds.) ICEC 2006. LNCS, vol. 4161, pp. 298–301. Springer, Heidelberg (2006). https://doi.org/10.1007/11872320_36

A Knowledge Acquisition Method for Event Extraction and Coding Based on Deep Patterns

Alfred Krzywicki[1]([✉]), Wayne Wobcke[1], Michael Bain[1], Susanne Schmeidl[2], and Bradford Heap[1]

[1] School of Computer Science and Engineering,
University of New South Wales, Sydney, NSW 2052, Australia
{alfredk,w.wobcke,m.bain,b.heap}@unsw.edu.au
[2] School of Social Sciences,
University of New South Wales, Sydney, NSW 2052, Australia
s.schmeidl@unsw.edu.au

Abstract. A major problem in the field of peace and conflict studies is to extract events from a variety of news sources. The events need to be coded with an event type and annotated with entities from a domain specific ontology for future retrieval and analysis. The problem is dynamic in nature, characterised by new or changing groups and targets, and the emergence of new types of events. A number of automated event extraction systems exist that detect thousands of events on a daily basis. The resulting datasets, however, lack sufficient coverage of specific domains and suffer from too many duplicated and irrelevant events. Therefore expert event coding and validation is required to ensure sufficient quality and coverage of a conflict. We propose a new framework for semi-automatic rule-based event extraction and coding based on the use of deep syntactic-semantic patterns created from normal user input to an event annotation system. The method is implemented in a prototype Event Coding Assistant that processes news articles to suggest relevant events to a user who can correct or accept the suggestions. Over time as a knowledge base of patterns is built, event extraction accuracy improves and, as shown by analysis of system logs, the workload of the user is decreased.

Keywords: Knowledge acquisition
Natural language and speech processing

1 Introduction

Identifying and tracking conflicts in order to anticipate the escalation of violence has been an elusive challenge for social science. The field of peace and conflict studies suffers greatly from the lack of reliable quantitative data on events which are more complex than the basic information typically collected for armed conflicts, such as individual battle statistics. This limits the ability of analysts to develop effective and general conflict analyses and early warning models [2].

© Springer Nature Switzerland AG 2018
K. Yoshida and M. Lee (Eds.): PKAW 2018, LNAI 11016, pp. 16–31, 2018.
https://doi.org/10.1007/978-3-319-97289-3_2

The idea of *coding* the data is to *annotate* relevant information with concepts from an ontology [11,20] defined by the analyst. In this paper, we focus on extracting events from news sources that provide timely and useful information on a daily basis.

The basic problem of event extraction is to convert unstructured text into an event structure whose main elements are actors (the "who"), action (the "what has been done") and targets (the "to whom"). Event coding also requires assigning to each event an *event type*, here associated with a number of event *categories* from an ontology, using a "codebook" that outlines the intended meaning/usage of the categories. Existing consolidated datasets lack sufficient coverage of specific domains and suffer from too many duplicated and irrelevant events [18,24]. Despite these numerous automatic extraction systems [10,12], human coding and validation of events is required to ensure sufficient quality and coverage [20].

In this paper, we presents a new framework for event extraction and coding based on the use of automatically generated *deep patterns* created during ordinary usage of the system by analysts, enabling incremental acquisition of a pattern knowledge base and improvement of the event coding process. Deep syntactic-semantic patterns derived from full sentence parse trees are able to locate the exact positions of event elements in more complex sentence structures, increasing linguistic coverage and accuracy, in contrast to "shallow parsing" methods typically based on identifying the main phrases occurring in a sentence [22]. Our incremental event extraction methodology is close in spirit to Ripple Down Rules (RDR), which has been applied to extracting relations [13] and ontologies [15] from text. However, since our knowledge acquisition takes the form of complex sentence patterns, we do not apply rule refinement as in RDR, rather we build up a large collection of patterns which are used heuristically to suggest potential events to a user.

As a demonstrator of the approach, we implemented a prototype Event Coding Assistant that enables a user to extract and code events from variety of news sources with concepts (event categories and entities) from an ontology. We focus on events from the long-running war in Afghanistan, reported in a variety of English language news sources (which we call the *AfPak* domain, following common usage of this term to denote Afghanistan/Pakistan).

Another key feature of our work is that we address the dynamic nature of the ontology, allowing users to modify the ontology and adapt to different language styles, by adding new English variants of the names of individuals, new entities such as emerging groups or group leaders (e.g. ISIS/Daesh), and new event categories. As an extreme example of different language style, in news reported by the Taliban web site (in English), the Afghan National Security Forces are called "puppets" or "hirelings", and their own casualties are described as "martyrs".

Our main research hypothesis is that automatically constructed deep patterns derived from ordinary user interactions are an effective way to extract and code domain specific events. We demonstrate this from an analysis of user behaviour over a 38-day trial of annotators using our Event Coding Assistant in the AfPak domain. We show by examination of system logs that the time taken for an annotator to code an event decreases with use of the system due to the reuse of deep patterns. Most importantly, the saving is achieved without any additional effort required from the users, and in particular they never need to view or edit a parse tree, and do not require knowledge of syntax.

The rest of the paper is organised in three main parts: the description the event extraction and coding framework, including the definition and application of deep patterns for event extraction and the representation of events, the description of the event coding interface, and the evaluation based on analysis of system logs. Following this is a discussion of related work.

2   Event Extraction and Coding Framework

In this section we present our event mining framework based on deep patterns, unification-based grammar formalisms and domain specific ontologies, illustrating this with reference to the AfPak domain. We start with an overview of the ontology.

2.1   AfPak Ontology

Generally, coding ontologies are hierarchical and information can be classified at different levels of the hierarchy. Ontologies for event categories usually take the form of a type hierarchy. As our objective is to code events relating to the conflict in Afghanistan, we adapted the basic CAMEO ontology [5], focusing on violent events (attacks, killings, etc.) and statements (public announcements, claims of responsibility, etc.).

In conjunction with a human expert, we defined an ontology with 210 event categories in three levels of hierarchy, with 87 concepts in the first level (e.g. Condemn), 119 in the second (e.g. Condemn.Killing civilians) and 4 concepts in the third level (e.g. Condemn.Killing civilians.Women and children). Under this coding scheme, events can be classified at any level, e.g. if an event is an attack but not one of the subtypes of attack specified in the hierarchy. Note that some of these concepts are domain specific due to the nature of the conflict (such as suicide bombing), though the aim is make the ontology as domain independent as possible. Overall, our ontology has 20 out of 87 first level categories closely corresponding to the main categories in CAMEO. As CAMEO is intended for much broader application in inter-state relations, we do not need many of the first level event categories.

The AfPak ontology also includes an extensive collection of domain specific concepts built from publicly available sources: 1521 concepts organised into five separate hierarchies (Table 1): Event Category, Person, Organisation, Location and Equipment. Some concepts correspond to entities, i.e. particular individuals (or roles of individuals), organisations or locations. Each entity has an associated canonical name and a list of alternative names.

Table 1. AfPak ontology with examples

Top level concepts	Count	Example
Event category	210	Condemn
Person	441	Abdul Rashid Dostum
Organisation	155	Taliban
Location	708	Kabul
Equipment	7	RPG

To facilitate semantic search, the ontology also shows basic relationships between concepts, such as "kind-of", "part-of" and "member-of". To give a few examples of such ontology triples, "RPG *kind-of* Weapon", "Argo (district) *part-of* Badakhshan (province)", "Minister of Defense *member-of* Government". For efficient querying, the AfPak ontology is implemented in the PostgreSQL relational database with a single concept in each row and relation names represented in columns.

2.2   Event Definition and Structure

An *event* is a structured object with one event type and a number of *components* (actors, targets and locations). However, because ontology concepts do not completely capture the meaning of a sentence expressing an event (so are best thought of as "annotations" of the sentence), an event structure also includes the lexical items (words) making up the event type and each actor, target and location.

More formally, an event is represented as an *event structure*, a lexical-semantic feature structure, with a *LEX* feature containing the string of lexemes (words) making up an event description, and a semantic frame consisting of an event *type* (an action phrase and a nonempty set of event *categories* from the ontology), and sets of feature structures for the *actors* and/or *targets* of the event (an event must have at least one actor or target), and (optionally) a set of *locations*. The feature structure for each actor, target and location contains the string of words making up the event component and any associated concept assigned from the ontology. An example event structure is as follows:

[*LEX*: *The Taliban killed five tourists*,
   EVENT: [*TYPE*: [*LEX*: *killed, Categories*: {*Action.Kill*}],
      ACTORS: {[*LEX*: *Taliban, Entity*: *Organisation.Taliban*]},
      TARGETS: {[*LEX*: *five tourists, Entity*: *Person.Civilians*]}]]

2.3   Pattern Creation and Generalisation

Pattern based event extraction forms the basis of our event mining framework. Patterns are created and used without the direct involvement of the annotator, from the segments of text highlighted during annotation.

1. If an NP node generates w, and all words in w except for a leading DT (determiner) or CD (cardinal number) are tagged with the same semantic role tag T, tag the DT or CD word with T.
2. Tag each preterminal node with the same semantic role tag (if any) of the word that it generates.
3. Take the minimal subtree of the parse tree whose leaf nodes n generate a string of words all with the same semantic role tag T, *and* which contains all nodes m with the same root path as any such node n.
4. For any leaf nodes n of this tree that generate strings of words that have the same semantic role tag T, assign a feature structure [T: V] where V is a new variable associated with T (a particular type, actor or target).
5. For any other node n, assign a feature structure that (i) if child nodes of n have the structure [ACTOR: V] or [TARGET: V], includes [ACTORS: {V}] or [TARGETS: {V}]; (ii) otherwise is [EVENT: [TYPE: E], [ACTORS: V1], [TARGETS: V2]] where E is the TYPE and V1/V2 are constrained to be the union of the ACTORS/TARGETS of the child nodes.

Fig. 1. Deep pattern creation

A *deep pattern* is a syntactic-semantic pattern in the spirit of unification-based grammar formalisms [23], that consists of a partial parse tree with feature structures assigned to (some of) the leaf nodes in the tree. Thus the deep patterns used in this paper are of a particularly simple form suitable for automatic generation. The feature structure for a leaf node has a *LEX* feature and a semantic feature that is either the event *TYPE* or an event component (*ACTOR* or *TARGET*). In the pattern, the values of these features are placeholder variables that will be instantiated when the pattern is applied to a new sentence.

A full parse tree for a sentence is derived using the OpenNLP[1] parser, and consists of a set of nodes that have a syntactic category, which for a preterminal node is a part of speech (POS) tag. The leaf nodes of the tree (terminal symbols of the grammar) are simply the words of the sentence. Trees may be written in list notation: e.g. the parse tree of "John hit the ball" is **(S (NNP John) (VP (VBD hit) (NP (DT the) (NN ball))))**. A node n *generates* the string of words w in a parse tree if w is the concatenation of the words in the subtree with root node n (thus the syntactic category of n generates w in the context-free grammar sense). The *root path* of a node n is the sequence of syntactic categories of the nodes on the path in the tree from n to the root, e.g. the root path of the VBD node is [VBD, VP, S].

The process of pattern creation begins with an *annotated sentence*, a sentence where some sequences of words have been *highlighted* and *tagged* with a semantic role (event type, actor or target). Note that a sentence can contain more than one sequence of words tagged with the same semantic role but with different tags (e.g. when a sentence describes an event with two or more targets and the target tags are distinct). The pattern creation process is defined in Fig. 1.

---

[1] https://opennlp.apache.org/.

We now illustrate this process with an example. Suppose the annotator has highlighted the text shown in the following sentence, where yellow denotes actors, red an event type and green a target of an event (again, note that the annotator selects only the text and does not see the parse tree):

(S (NP (NNP Taliban) (NNS insurgents)) (VP (VBP have) (VP (VBN shot) (ADVP (RB dead)) (NP (DT a) (JJ 19-year-old) (JJ Sar-e-Pul) (NN woman)))))

In Step 1, the untagged DT "a" in the target NP "a 19-year-old Sar-e-Pul woman" is tagged with the same target tag. In Step 2, the tags are propagated from the terminal (word) nodes to the preterminal nodes. To show this, we colour the preterminal nodes:

(S (NP (NNP Taliban) (NNS insurgents)) (VP (VBP have) (VP (VBN shot) (ADVP (RB dead)) (NP (DT a) (JJ 19-year-old) (JJ Sar-e-Pul) (NN woman)))))

In Step 3, the minimal subtree of this tree is found that generates all the tagged words and contains as leaf nodes all nodes with the same root path as one of these nodes. In this case, the second condition is satisfied and in effect the parse tree is pruned so as to generate only the tagged text. In Step 4, the leaf nodes are assigned a feature structure that has the semantic role attribute with a value that is a new variable:

(S NP[ACTOR: A] (VP (VP VBN[TYPE: E] NP[TARGET: T])))

The final step assigns feature structures to non-leaf nodes in this tree. The VP node is assigned the structure [EVENT: [TYPE: E, TARGETS: {T}]. The S node is assigned the final semantic frame of the event structure: [EVENT: [TYPE: E, ACTORS: {A}, TARGETS: {T}].

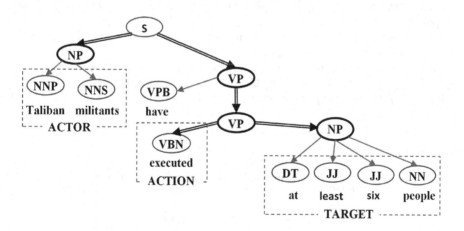

Fig. 2. Pattern application and event extraction

2.4   Pattern Based Event Extraction

Suppose a new sentence has been given a full parse tree. We now define how a pattern matches a parse tree and can be used to extract an event. First note that a pattern always has a root node with category S, but the pattern can be matched to any subtree of the sentence parse tree that also has a root node S. A pattern P matches a tree T if they have the same root, and considering the syntactic categories of the nodes only, P is a subtree of T. A pattern P could match a tree in multiple ways: for any such subtree T, we take only the "leftmost" such match. More precisely, suppose a pattern matches two subtrees T_1 and T_2 of the sentence parse tree. T_1 is to the *left* of T_2 if the first node of the sentence parse tree where the subtrees T_1 and T_2 differ in a depth-first traversal of the parse tree is earlier in the traversal in T_1 than in T_2.

Each node p in the chosen pattern matches a node n in the sentence parse tree. Feature structures are now assigned to nodes in the parse tree. Suppose p is a leaf node of the pattern so is assigned a feature structure [T: V]. Entities and event categories are found by matching the text in the sentence directly to the ontology. Then n is assigned a feature structure with (i) a *LEX* attribute whose value is the string of words generated by n, and (ii) an *Entity* attribute whose value is the ontology entity annotating this string of words, except in the case of the event type, where the set of ontology entities becomes the value of the *Categories* attribute. The variable V is instantiated to this feature structure. Following that, values propagate up the tree such that the *LEX* values of parent nodes inside an event structure are the concatenation of the *LEX* values of child nodes in the event structure. In this way, the *LEX* attribute values and the entities assigned to variables come from the sentence, while the semantic frame for the event that includes the event type and components comes from the pattern.

For example, the pattern created in the previous section, when applied to the following sentence (see Fig. 2):
(TOP (S (NP (NNP Taliban) (NNS militants)) (VP (VBP have) (VP (VBN executed) (NP (QP (IN at) (JJS least) (CD six)) (NNS people))))))
assigns [*ACTOR*: [*LEX*: Taliban militants, Entity: Organisation.Taliban]], [*Type*: [*LEX*: executed, Categories: {Kill.Execute}]] and [*TARGET*: [*LEX*: at least six people, Entity: Person.Civilians]] to the NP, VBN and second VP nodes. The event structure assigned to the S node is as follows:

$$
\begin{bmatrix}
LEX : \text{Taliban militants executed at least six people} \\
EVENT : \\
\quad \begin{bmatrix}
TYPE : \begin{bmatrix} LEX : executed \\ Categories : \{Kill.Execute\} \end{bmatrix} \\
ACTORS : \left\{ \begin{bmatrix} LEX : \text{Taliban militants} \\ Entity : Organisation.Taliban \end{bmatrix} \right\} \\
TARGETS : \left\{ \begin{bmatrix} LEX : \text{at least six people} \\ Entity : Person.Civilians \end{bmatrix} \right\}
\end{bmatrix}
\end{bmatrix}
$$

3   Event Coding Assistant

In this section, we present the main interface to an Event Coding Assistant and describe its operation through a typical workflow for coding a single document. Figure 3 outlines this workflow, separately including a Use Case containing the steps performed by the user and the system at various points. Note that this is a standardised workflow: users can also move between events, search the ontology, etc., so are not restricted to following this particular series of steps.

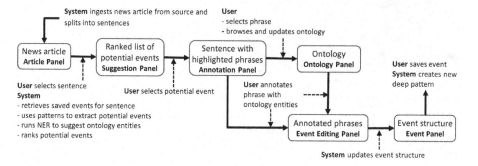

Fig. 3. Annotation and coding workflow

Prior to this workflow commencing, there is a pre-processing step in which news articles are ingested from their source on the Web and any source formatting and images are removed. The article is then automatically split into sentences that are parsed into a tree structure.

A simple filter is run to check whether the article contains any string matching any of the list of names associated with any entity in the ontology. If an article contains at least one such string, it is displayed in the Article Panel (left top panel in Fig. 4) for user consideration.

The first sentence is selected by default and the user can select any sentence in the article within the Article Panel. The system retrieves and displays any stored events for the sentence from previous coding and marks it in blue. If there are no stored events for this sentence, event suggestions are immediately displayed for the selected sentence in the Suggestion Panel (bottom left panel in the figure).

The suggestions are computed by applying deep patterns to the parse tree for the sentence, enabling the system to identify phrases for actors, event type (action) and targets as described in Sect. 2.4. The first step is to run named entity recognition over the sentence, based on keyword matching of text strings to the list of canonical and alternative names of entities in the ontology. As more than one event can be extracted from a sentence, the potential events are displayed as a ranked list based on scoring which takes into account the presence of event components, the number of ontology entities and the specificity of event type string. The user can then select a potential event that most closely matches

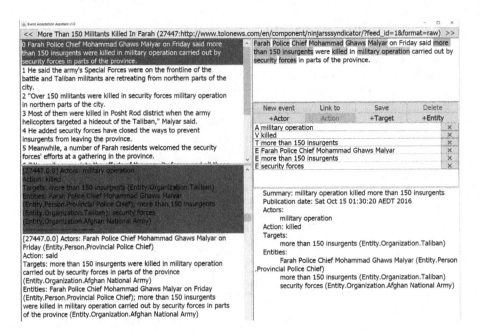

Fig. 4. Interface for annotating events (Colour figure online)

the expected structure. Finally, the system displays the selected potential event in the Annotation Panel (the top right panel of the interface), highlighting the actor, action and target phrases with yellow, red and green colours respectively. Other phrases that have been matched with ontology entities are highlighted in blue and displayed in the Event Editing Panel (right middle). The resulting event structure is shown in the Event Panel (right bottom).

Any required editing is made using both the Event Annotation and Event Editing panels. In the Event Annotation Panel, the user can unhighlight phrase keywords and highlight others until the phrases are correct. The changes are immediately shown in the Event Editing Panel, where the user has the option to delete any phrase or add a new phrase for an actor, action or target. The system also allows the user to assign an ontology entity to an arbitrarily selected phrase if the phrase is not pre-highlighted in blue. Before saving the event, the user is required to add one or more event categories by clicking on the category in the Ontology Panel.

After editing is complete, the event, together with any newly generated pattern, is saved, which is reflected by the system displaying the saved event, as shown in Fig. 5. At this point the user can proceed to the next article or return to editing previously saved events. At any time, the user has the option to add an entity to the ontology, such as a new event type or organisation, using the Ontology Panel (Fig. 6), or add an alternative name to an existing entity. This is very useful as the domain is highly dynamic in nature and it is impossible to construct a complete ontology in advance.

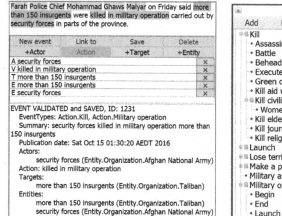

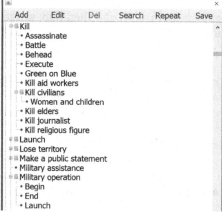

Fig. 5. Final event   **Fig. 6.** Ontology panel

4   Evaluation

In this section, we evaluate the Event Coding Assistant, showing that the system reduces event annotation time due to the use of deep patterns, learnt from user interaction through normal usage of the system. We also show that the number of reused patterns grows over time. Our secondary goal is to assess the usability of the Assistant and observe how users interacted with the system. Our observations and measurements are based on three sources on information derived from a 38-day trial of the system: the event and pattern database, the system log with recorded user actions, and a user questionnaire administered after the event annotation was completed. We begin by describing the evaluation of event coding and the dataset created by the annotators, including the steps taken to ensure that the annotators were performing the event coding task adequately.

4.1   Evaluation of Event Coding

The Event Coding Assistant was evaluated by two annotators, one male and one female, who used the system for 38 days over a three month period covering December 2016 to February 2017. In total, the annotators read 7517 news articles, of which 826 contained relevant events. From these articles, the annotators coded 1478 events, from which the system generated 1298 deep patterns. The analysis of event creation and pattern usage in this section is based only on the 1361 events and 1175 patterns that were defined after an introductory pilot of the system by users other than the annotators. The events in the analysis do not include events created and then later deleted.

The annotators were social science graduates who had studied international relations, but they had no experience in event coding, no specialist knowledge of the AfPak domain, and had not used any event annotation tool before.

Annotator 1 worked alone for the first 7 days, followed by Annotator 2 working for the next 15 days, and then they worked concurrently for the final 16 days. The annotators were instructed to consistently and completely cover all significant events.

To evaluate the user perception of the Event Coding Assistant, annotators filled in a questionnaire after the end of the trial. The annotators commented that the interface was relatively easy to understand, learn, navigate and define events. Suggestions were used by both annotators and, in their opinion, improved over time. Most of the time the annotators used the first four suggestions from the ranked list.

After the annotation was completed, about 10% of events were validated by an expert social scientist. Out of 150 events validated, 79% were confirmed as accurate by the expert. Of the 136 event categories used, the majority of events from all sources were categorised as military operations involving casualties, followed by some kind of terrorist attack. The main reason for this distribution is that these events were prominent in news articles. The least common categories were about making public statements, claims and protests.

Table 2. Event dataset characteristics

Event structure	Count	Percent
Events with actors, action and targets	444	33%
Events with more than one type	642	47%
Events with more than one actor	37	2.7%
Events with more than one target	32	2.3%
Events with no actors	559	41%
Events with no targets	592	43%

Table 2 gives counts of events with different characteristics. Only about 33% of events contain actors, action and targets. The allocation of events into categories is also a typical long tail distribution, with 15% of events in the top "Kill" category, with frequencies of categories decreasing rapidly into a long tail. More than half the events have only one category, and the mean over the whole set is 1.6 categories per event.

4.2   Evaluation of Pattern Generalisation

To assess how deep pattern generation supported the annotators' work, we analysed the annotation log, extracting the number of saved events per day and the number of user actions performed to create and edit events for each of the 38 annotation days. The time estimates exclude time spent on browsing past articles and prior events, and other activities such as reading the event coding guidelines, short breaks, etc., however, the estimates do include time spent on updating an event after it was created.

Figure 7 shows mean annotation time per *saved event*, split into the mean time spent on creating events and that spent on updating events. The time per event is calculated by dividing the total annotation time for each day by the number of saved events for that day. This measures the effective user effort to create all the saved events for the day: note that this includes editing events that were not saved to the event database. Figure 7 shows clearly that this mean time per saved event decreases over the course of the 38 days, from a peak of around 120 s to around 50 s, with the mean time spent on creating and updating events also decreasing.

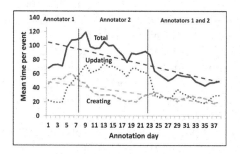

Fig. 7. Event coding time (Colour figure online)

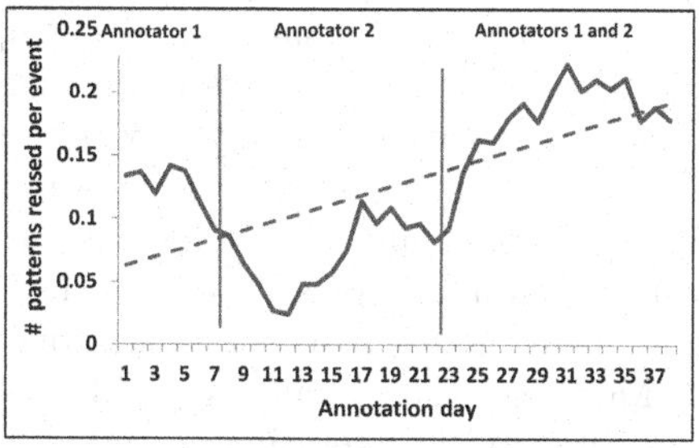

Fig. 8. Pattern reuse (Colour figure online)

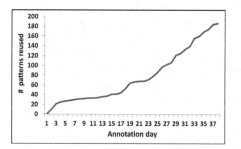

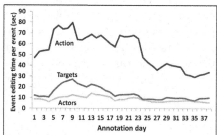

Fig. 9. Cumulative pattern reuse (Colour figure online)

Fig. 10. Editing time per event component (Colour figure online)

We now show that the diminishing time per saved event can to a large extent be attributed to deep pattern learning from user actions. A new deep pattern is created from a new or updated event only when the final event does not match any existing pattern. If a pattern is reused in an event, the event structure is correct and only minimal work is required from the analyst to review and save the event. Therefore it is desirable that a high number of patterns are reused. Figure 8 plots pattern reuse per created event, clearly showing that pattern reuse increases over time. Pattern reuse begins to rise faster around day 23, just when the time taken per saved event starts to drop more rapidly back to below 100 s.

We can make some general observations on the trends in Figs. 7 and 8. Both plots in these figures are divided into three periods, corresponding to the three periods of annotation described in Sect. 4.1. Each of these periods has its own annotation style. In the short, 7-day period labelled as "Annotator 1", we observe, after a few days of stable usage, an increase in the time taken per event and a decrease in pattern reuse. During this period, the update part of event times started to increase (finely dashed green line) and the creation time of events decreased (coarsely dashed red line). This trend continues into the second "Annotator 2" period. After Day 11 however, the reuse of patterns starts to pick up until another decline after Day 17. These ups and downs in pattern reuse correspond inversely to changes in event update time, suggesting that Annotator 2 quickly created many events using suggestions, then revised these events which caused the creation of many new patterns. The "Annotators 1 and 2" period is very different in that we observe a rapid decline in update time, decrease in total time per event, and steady increase in pattern reuse. This period is where we believe the annotation style and the usage of the system started to be as expected and where the reduction of annotators workload is visible.

Moreover, the cumulative number of reused patterns grew substantially over time (Fig. 9). The top five patterns, used from from 10 to 20 times, are very simple, extracting only actors and action or action and targets. For example, the pattern **(S NP[TARGET: T] (VP VBD[TYPE: V] (VP VBN[TYPE: V] (PP IN[TYPE: V] NP[TYPE: V]))))** was used 20 times and extracted events such as "Five Daesh fighters were killed in an airstrike". Another pattern **(S NP[ACTOR: A] (VP VBD[TYPE: V] NP[TARGET: T]))** was reused 10 times to extract events such as "Afghan government executed 6 terrorists".

Figure 10 shows the breakdown of editing time per event into actors, actions and targets. The action component of the event takes around three times longer than actors and targets, presumably because this requires identifying an action phrase and browsing the ontology for one or more event categories, whereas actors and targets are usually identified by named entity recognition. This figure also shows a substantial reduction in time spent identifying the event type over the course of the trial. Targets generally take more time to edit than actors, which agrees with the annotators' comments.

5   Related Work

In this section, we briefly summarise recent work related to event extraction and annotation, both manual and automatic. Event extraction systems have a long history starting in the 1990s. More recent work typically involves the use of machine learning combined with NLP parsing. To this end, methods can be divided into feature based with various levels of supervision [4,8,17], and pattern based [6,19]. Event extraction methods can use an aggregation of documents [4,6,8] or single documents, or process either a whole document or sentences one at a time, e.g. PETRARCH [19].

JET and PETRARCH can be considered related to our research as they both use extraction patterns. JET [6] uses pattern matching to identify trigger words and event types, then statistical modelling to identify and classify the arguments, roles and relevance of populated events. For every "event mention" in the ACE training corpus, patterns are constructed based on the sequences of constituent heads separating the trigger and arguments. In our system, patterns identify arguments and their roles in the sentence structure, mapped to the ontology in the second stage of processing. An extension of JET [21] captures the relationship between arguments to ascertain if they both belong to the same or different events. We are not aware of any use of this system or its extensions outside of the ACE framework. The TABARI/PETRARCH method of extracting events [1] uses a large set (over 10,000) handcrafted and tuned verb patterns.

To differentiate from our work and for completeness, we mention related research applying deep parsing for triple extraction [16] and extracting biological events [3]. The former searches the sentence parse tree for subject-predicate-object relations, where each of these elements is found separately in either a pre-defined place in the tree structure, or a place determined by heuristic rules. The latter uses just two simple manually constructed deep patterns for finding biological events in two types of sentence structures. In contrast, we are dealing with very complex sentences for which automatically generated patterns find event elements in varying positions of the sentence tree, and event extraction is guided by a domain specific ontology.

There are several systems in the social sciences that make use of complex patterns to extract events [10,19], however these systems typically use a large number of manually constructed patterns (in TABARI over 10000) that are highly specific [1]. Constructing these systems is extremely time consuming because the sets of patterns are labour intensive to create, and then very difficult to maintain. Generally, automated event extraction systems without human involvement report duplicated stories and are prone to missing important events. This may not be a problem for consolidated analysis, but is unacceptable when the focus is on a specific domain as it may lead to making wrong decisions or missing necessary action [7]. A human annotator, on the other hand, can apply their external contextual knowledge, increasing the relevance of coded events, but requires extensive training of the coding process to ensure coding consistency [11]. Nevertheless, despite the existence of fully automated coding systems, such as GDELT [10] and ICEWS [12], many other coding projects in the social sciences, such as the Global Terrorism Database (GTD) [9] and the Armed Conflict Location and Event Data Project (ACLED) [14], remain human based to ensure high accuracy.

Nardulli et al. [11] present a hybrid, machine and human, approach to event coding, called SPEED (Social, Political and Economic Event Database). The motivation is that machines are good at extracting "manifest" textual content, such as word count, co-occurrence and position, whereas humans cannot be surpassed on the "latent" content, requiring judgement and background knowledge. The SPEED hybrid system uses a pre-trained Naive Bayes model to classify a portion of input documents into pre-defined topics, leaving the rest out as irrelevant. Initially the SPEED system had high (98%) recall and low (33%) accuracy.

To correct this, extra training was provided using an additional 66,000 human processed documents. This increased the accuracy to 87% while maintaining high recall. The authors used the term "progressive" to describe the improvement of the machine learning algorithms from human feedback. However, it is not clear how this is done based on the batch learning example given in the paper.

6   Conclusions and Future Research

We proposed a knowledge acquisition framework that can be used to extract events contained in a news articles and suggest how those events should be coded using a domain specific ontology for later search and analysis. This is achieved by progressive generation of patterns from deep parsing of sentences and incremental coding of events by a human annotator.

We described a prototype Event Coding Assistant to facilitate event coding in the Afghanistan/Pakistan domain. The prototype uses an extensive domain specific ontology of event categories and individuals, organisations and locations. By examining system logs of a 38-day trial of the system using two annotators, we showed that the time taken for an annotator to code an event decreases with time due to the reuse of deep patterns automatically generated from normal user interaction with the system.

The Event Coding Assistant adapts to the language used to express events in a variety of news sources using the entities and event types specific to the domain. Thus our approach provides an effective way to develop systems for coding domain specific events with high accuracy and for lower cost than systems based on a large number of manually constructed patterns. For future work, we plan to improve event categorisation and ranking using machine learning methods, and further study stream mining techniques for event extraction and coding.

Acknowledgements. This work was supported by Data to Decisions Cooperative Research Centre. We are grateful to Michael Burnside and Kaitlyn Hedditch for coding the AfPak event data.

References

1. Best, R.H., Carpino, C., Crescenzi, M.J.: An analysis of the TABARI coding system. Confl. Manag. Peace Sci. **30**, 335–348 (2013)
2. Bond, D., Bond, J., Oh, C., Jenkins, J.C., Taylor, C.L.: Integrated data for events analysis (IDEA): an event typology for automated events data development. J. Peace Res. **40**, 733–745 (2003)
3. Bui, Q.C., Sloot, P.: Extracting biological events from text using simple syntactic patterns. In: Proceedings of the BioNLP Shared Task 2011 Workshop, pp. 143–146 (2011)
4. Fung, G.P.C., Yu, J.X., Yu, P.S., Lu, H.: Parameter free bursty events detection in text streams. In: Proceedings of the 31st International Conference on Very Large Data Bases, pp. 181–192 (2005)

5. Gerner, D.J., Schrodt, P.A., Yilmaz, O., Abu-Jabr, R.: Conflict and mediation event observations (CAMEO): a new event data framework for the analysis of foreign policy interactions. In: The Annual Meetings of the International Studies Association, New Orleans, LA (2002)
6. Ji, H., Grishman, R.: Refining event extraction through cross-document inference. In: Proceedings of the 46th Annual Meeting of the Association for Computational Linguistics, pp. 254–262 (2008)
7. Kennedy, R.: Making useful conflict predictions. J. Peace Res. **52**, 649–664 (2015)
8. Kuzey, E., Vreeken, J., Weikum, G.: A fresh look on knowledge bases: distilling named events from news. In: Proceedings of the 23rd ACM International Conference on Information and Knowledge Management, pp. 1689–1698 (2014)
9. LaFree, G., Dugan, L.: Introducing the global terrorism database. Terrorism Polit. Violence **19**, 181–204 (2007)
10. Leetaru, K., Schrodt, P.A.: GDELT: global data on events, location, and tone, 1979–2012. In: The Annual Meetings of the International Studies Association, San Francisco, CA (2013)
11. Nardulli, P.F., Althaus, S.L., Hayes, M.: A progressive supervised-learning approach to generating rich civil strife data. Sociol. Methodol. **45**, 148–183 (2015)
12. O'Brien, S.P.: Crisis early warning and decision support: contemporary approaches and thoughts on future research. Int. Stud. Rev. **12**, 87–104 (2010)
13. Pham, S.B., Hoffmann, A.: Incremental knowledge acquisition for extracting temporal relations. In: Proceedings of the 2005 12th IEEE International Conference on Natural Language Processing and Knowledge Engineering, pp. 354–359 (2005)
14. Raleigh, C., Linke, A., Hegre, H., Karlsen, J.: Introducing ACLED: an armed conflict location and event dataset. J. Peace Res. **47**, 651–660 (2010)
15. Ruiz-Sánchez, J.M., Valencia-García, R., Fernández-Breis, J.T., Martínez-Béjar, R., Compton, P.: An approach for incremental knowledge acquisition from text. Expert Syst. Appl. **25**, 77–86 (2003)
16. Rusu, D., Dali, L., Fortuna, B., Grobelnik, M., Mladenić, D.: Triplet extraction from sentences. In: Proceedings of the 10th International Multiconference Information Society - IS 2008, pp. 8–12 (2007)
17. Rusu, D., Hodson, J., Kimball, A.: Unsupervised techniques for extracting and clustering complex events in news. In: Proceedings of the Second Workshop on EVENTS: Definition, Detection, Coreference and Representation, pp. 26–34 (2014)
18. Schrodt, P.A.: Automated production of high-volume, real-time political event data. In: APSA 2010 Annual Meeting Papers (2010)
19. Schrodt, P.A., Beieler, J., Idris, M.: Three's a charm?: open event data coding with EL:DIABLO, PETRARCH, and the open event data alliance. In: The Annual Meetings of the International Studies Association, Toronto, ON (2014)
20. Schrodt, P.A., Yonamine, J.E.: A guide to event data: past, present, and future. All Azimuth **2**(2), 5–22 (2013)
21. Sha, L., Liu, J., Lin, C.Y., Li, S., Chang, B., Sui, Z.: RBPB: regularization-based pattern balancing method for event extraction. In: Proceedings of the 54th Annual Meeting of the Association for Computational Linguistics, pp. 1224–1234 (2016)
22. Shellman, S.M.: Coding disaggregated intrastate conflict: machine processing the behavior of substate actors over time and space. Polit. Anal. **16**, 464–477 (2008)
23. Shieber, S.M.: An Introduction to Unification-Based Approaches to Grammar. CSLI Publications, Stanford, CA (1986)
24. Ward, M.D., Beger, A., Cutler, J., Dickenson, M., Dorff, C., Radford, B.: Comparing GDELT and ICEWS event data. Analysis **21**, 267–297 (2013)

Incremental Acquisition of Values to Deal with Cybersecurity Ethical Dilemmas

Deborah Richards[1(✉)], Virginia Dignum[2], Malcolm Ryan[1], and Michael Hitchens[1]

[1] Macquarie University, Sydney, NSW 2109, Australia
{deborah.richards,malcolm.ryan,
michael.hitchens}@mq.edu.au
[2] Delft University of Technology, Delft, The Netherlands
m.v.dignum@tudelft.nl

Abstract. Cybersecurity is a growing concern for organisations. Decision-making concerning responses to threats will involve making choices from a number of possible options. The choice made will often depend on the values held by the organisation and/or the individual/s making the decisions. To address the issue of how to capture ethical dilemmas and the values associated with the choices they raise, we propose the use of the Ripple Down Rules (RDR) incremental knowledge acquisition method. We provide an example of how the RDR knowledge can be acquired in the context of a value tree and then translated into a goal-plan tree that can be used by a BDI agent towards supporting the creation of ethical dilemmas that could be used for what-if analysis or training. We also discuss the AORTA framework can be extended with values to allow the BDI cognitive agent to take into consideration organisational norms and policies in its decision-making.

Keywords: Beliefs desires intentions · Agents · Values · Ripple Down Rules AORTA · Cybersecurity

1 Introduction

Cybersecurity is a growing concern to organisations, society, nations and globally. Central to the problem is a fundamental reliance on individuals to be aware, vigilant and responsible. There is increasing recognition that human factors are the cause of many cybersecurity breaches [1, 2]. In some cases ignorance is the problem, but often the issue is an individual's decision, or indecision, that results in organizational policy and procedures being ignored or compromised. This problem is not isolated to non-technical end-users or operational/low-ranking employees, but is pervasive across roles and levels of responsibility, including system administrators and those involved in designing software systems. The breaches often occur because the individual does not believe that the potential threat will happen to them and because their choices are often driven by personal values that rank non-compliant actions higher than actions that are compliant. Despite the unintended adverse ethical consequences of many breaches experienced by others and/or the individual, such as disclosure of private information,

K. Yoshida and M. Lee (Eds.): PKAW 2018, LNAI 11016, pp. 32–45, 2018.
https://doi.org/10.1007/978-3-319-97289-3_3

unethical choices are made because it is difficult for individuals to reason about the ramifications of what may appear to be a trivial or harmless decision or to comprehend how or why an incident might occur.

To prepare and educate employees, particularly managers, we are increasingly seeing organisations like PwC[1] and Ernst and Young[2] run cybersecurity training using simulations involving live role playing games or computer-based games. In some cases these online games use multimedia such as videos[3] to allow players to explore their options and the impact of their choices on the decision-making process and outcomes. Rather than using human actors or live role-playing that have time, space and perhaps skill limitations, we propose the use of intelligent virtual agents and virtual environments to help expose the individual's subconscious values, provide explanations about choices, policies and outcomes and identify individual risk areas. Furthermore, a computerised agent-based approach would allow "what-if analysis" to help organisations and individuals with risk planning.

Organisations implement technologies, policies and processes to manage cybersecurity issues. However, personal values drive humans to override norms in their decision-making [3, 4] resulting in policy breaches. To provide realistic ethical simulations for training and analysis, we want to create agents that are aware of organizational policy and also able to reason according to their values. We propose to use the AORTA framework [5] that allows a belief-desires-intention (BDI) [6] agent to reconcile its decision-making with organisational policy. We extend AORTA so that the agent is able to represent a human decision-maker and extend the BDI agent model to include human values, along the lines of Schwartz' value model [7] comprised of: self-direction, universalism, benevolence, conformity and tradition, security, power, achievement, hedonism and stimulation. We also intend to draw on the work of [8] to represent and reason about the effect of values on a BDI-agent's plan selection.

In this paper, we focus on a major impediment to achieving this goal. The well-known knowledge acquisition (KA) bottleneck. KA is the major hurdle we face in delivering a scalable approach to cybersecurity simulations that can apply to different organizations, roles, individuals and dilemmas that would allow individuals to explore their own beliefs, values and risk-levels relating to cybersecurity. The knowledge needed for such simulations include plausible scenarios, organizational policy and the agent's beliefs, goals, plans and values within a given context. Furthermore, being able to tailor the knowledge to specific organisations and individuals and provide ethical *contextual integrity* (i.e. different social spheres and cultural norms) requires an approach that supports knowledge maintenance and evolution. For this purpose, we turn to an incremental knowledge acquisition and representation method known as Ripple Down Rules [9].

The structure of the remaining paper is as follows. Section 2 presents the theoretical foundation of the research design. Section 3 explains in detail about the method of our

---

[1] https://www.pwc.com.au/cyber/game-of-threats.html.

[2] http://www.ey.com/Publication/vwLUAssets/EY_-_Cybersecurity_Incident_Simulation_Exercises/ $FILE/EY-cybersecurity-incident-simulation-exercises-scored.pdf.

[3] http://targetedattacks.trendmicro.com/.

analysis. We present our results in Sect. 4, followed by discussions of the results in Sect. 5. The paper ends with final conclusions in Sect. 6.

2 Background and Theoretical Foundations

In an European Union Horizon 2020 report, Christen, Gordijn, Weber, van de Poel and Yaghmaei [4] identified a number of ethical values relevant to cybersecurity including: social justice, equality, fairness, discrimination prevention, privacy, personal freedom, physical harm prevention, information harm prevention (see Fig. 1).

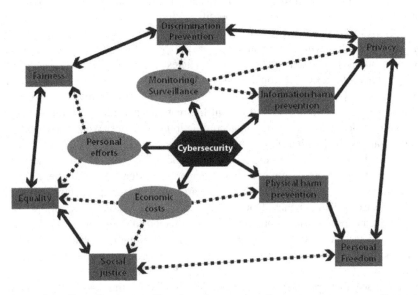

Fig. 1. Draft outline of value conflicts in cybersecurity. Orange squares are conflicts, solid arrows show positive (i.e., supporting) relations, dotted arrows show conflicting relations [4]

We see that some values are consistent and compatible with each other (solid line) but some values conflict with one another (shown by dotted lines). For example, privacy and personal freedom are compatible values that allow an individual to act as they choose without the knowledge or interference of others but personal freedom conflicts with the goals of social justice that is compatible with the values of equality and fairness. To support these goals will involve some actions/costs/effort (shown in blue). For example, to support social justice values will involve economic costs. Even though all these values are relevant for cybersecurity, in most realistic situations it is not possible to uphold all values simultaneously. This means that choices must be made concerning the prioritization of the values. Breaches of security are often a direct consequence of how values are prioritized. In the constraints-based implementation method in [8] that we propose to follow, two values that are in conflict (i.e. that pull in different directions) can not both be important and one is removed.

In this section we introduce the concepts, theories and approaches we aim to draw together to support creation of cybersecurity scenarios involving value-based BDI agents. To provide a theoretical foundation and guidance on identification of values for the cybersecurity context, we first present Schwartz's Values Theory, which we draw upon later in our example.

2.1 Schwartz's Values Theory

In contrast to culture-based theories, such as Hofstede's cultural dimensions [10], that look at differences between cultures, Schwartz [7] studied human values across approximately 60 cultures and concluded that a number of universal values existed in order for the society to meet individual and group physiological, coordination and survival needs. The compatibility of values with BDI agents is evident in the description of values that are comprised of beliefs and goals and influence plans and actions. However, one could see values as more salient in human decision making as value-driven goals motivate, value-driven beliefs evoke emotions and values (such as honesty) transcend norm driven actions, which might be specific to certain contexts. Values are ordered according to their relative importance by the individual and thus guide the individual's choices and their evaluation of situations, policies and others, helping them to prioritise the importance of and ultimately guiding the individual's actions.

There are four potentially competing groups (Openness to change, Self transcendence, Conservation, Self enhancement) encompassing10 value types have been identified as follows:

1. Self-Direction – extent to which someone prefers independent thinking and decision making.
2. Stimulation – importance of challenge, change and novelty in life.
3. Hedonism – the desire for self-gratification and positive experiences.
4. Achievement – role of competence, personal achievement and success.
5. Power – extent of control, recognition and influence over others.
6. Security – desire to minimize risk and uncertainty.
7. Tradition – relevance of observance of cultural values and customs.
8. Conformity – the importance of adhering to societal norms and expectations.
9. Benevolence – the importance of caring for those in your in-groups.
10. Universalism – the importance of caring for those outside your in-groups through ensuring equality, tolerance and mutual understanding.

Given the obvious relevance to cybersecurity we describe the security value type in more detail. As described by Schwartz [7] the defining goal of security is safety, harmony, and stability of society, of relationships, and of self. These values stem from the needs of individuals at the individual level (e.g. health, sense of belonging, cleanliness) and their needs that impact more widely on groups (e.g. social order, family security and national security). As an example, we can map the ethical values identified in [11] using Schwartz's value types. Social justice, equality, fairness and discrimination prevention fall under Universalism. Privacy and personal freedom fall under self-direction and power. Physical harm prevention and information harm prevention fall under conservation. Some values fall under multiple value types.

2.2 AORTA

AORTA [5] is a multi-agent system framework that allows agents to work together within the context of organizational policies and norms. The agents remain independent of the organization, following their own goals and plans, but are able to reason about specific organization concern (objectives, norms, or requirements). It is up to each agent to determine how it integrates organizational rules with its own reasoning rules. This enable a flexible and realistic approach to organizational behavior, as it can consider possible individual deviations. An organizational metamodel is used to map to specific organizational models which allows the agent to work in multiple organisations with varying policies.

Organizational reasoning in AORTA divided into three phases: obligation check (OC), option generation (OPG) and action execution (AE). The OPG-phase uses the organizational state to generate possible organizational options. The OC-phase uses the agent's mental state and organizational state to determine if obligations are activated, satisfied or violated, and updates the organizational state accordingly. The agent considers these options in the AE- phase using reasoning rules, which can alter the organizational state, the agent's intentions or send messages to other agents. The AORTA architecture is extended from [5] in Fig. 2, with an additional component to support values.

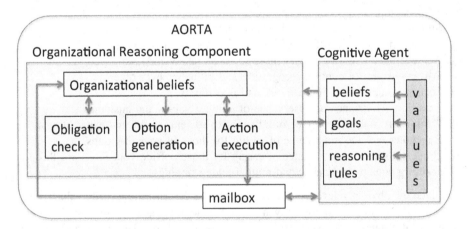

Fig. 2. AORTA architecture with values added to the cognitive agent

2.3   Adding Values to BDI Agents

As a first step to adding values to a BDI agent, we turn the abstract values defined by Schwartz into concrete values specific to that agent's role and/or the agent's decicion-making context that link to concrete agent goals. For this purpose we define a goal tree, G, as:

$$G ::= A \mid SEQ(G_{1-n}) \mid AND(G_{1-n}) \mid OR(O_{1-n})$$

where A is an action, or a sequence (SEQ) of subgoals, G_i; or a combination of subgoals in unspecified order (AND), or a set choices (OR) where each option $O_i =$ (C_i, G_i) define which condition can be selected for which subgoal. For $G :: = A$, the action is performed when the preconditions hold. For $G :: = SEQ(G_{1-n})$, all of the subgoals are executed in the specified order. For $G :: = AND(G_{1-n})$, the sub-goals are executed in an unspecified order. For $G :: = OR(O_{1-n})$, when condition Ci holds, the subgoal is executed when the associated option is chosen.

Values indicate the relative value or preference of multiple measurable criteria that are weighted according to the context. Thus the weightings are not fixed but depend on preferences, current situation and practicality. We annotate the nodes of the goal tree by annotating them with values in the form of weights to indicate their relative value, as described in [8]. Values could be the specific concrete values associated with cybersecurity, depicted in Fig. 1.

We draw on the Design for Values method [12] which involves creation of a value hierarchy is comprised of values (top layer), norms (middle) and design requirements (bottom layer). Each of these layers can also form a hierarchy or tree. In this paper we focus on values and thus we later present a value tree for a cybersecurity context.

2.4   Ripple Down Rules

Ripple Down Rules (RDR) uses cases to acquire rules directly from the domain expert, thus removing the need for a knowledge engineer to mediate in the knowledge acquisition (KA) process. Knowledge maintenance is facilitated through local patching of rules in the context of the case/s, known as cornerstone cases, associated with the rule being modified. Figure 3 provides a simple view of the knowledge acquisition process. Cases provide grounding and features are chosen from the case to provide the "justification" for the classification assigned. Multiple Classification RDR (MCRDR) is an extension of single classification is defined as the quadruple (rule, P, C, S), where P is the parent rule, C are the children/exception rules

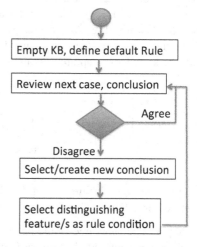

Fig. 3. RDR Standard KA process

and S are the sibling rules within the same level of decision list. Every rule in the first list is evaluated. If a rule evaluates to false then no further lists attached to that rule are examined. If a rule evaluates to true all rules in the next list are tested. The list of every true rule is processed in this way. The last true rule on each path constitutes the conclusions given. For the case {a,d,g,i,k}, the MCRDR tree in Fig. 4, three rules with two conclusions are given: Class 5 (Rule 7), Class 7 (Rule 8) and Class 7 (Rule 9).

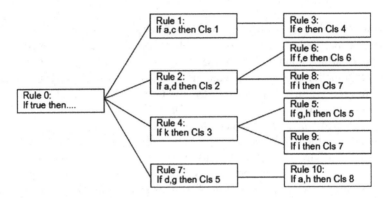

Fig. 4. An MCRDR knowledge base.

3 Proposed Approach

The overarching goal is to draw on the aforementioned concepts and approaches to support simulation of cybersecurity ethical dilemmas. To achieve this goal we intend to follow the process shown in Fig. 5. Working backwards from our goal in Fig. 5, to allow the BDI agents in our cybersecurity simulations to reason about values, we modify the cognitive agent in AORTA to include values that are used by the reasoning rules, as shown in Fig. 2. Continuing to work backwards in Fig. 5, the agent's goals, values and reasoning rules are obtained from the goal-plan tree that has been extracted from the knowledge captured via the modified RDR KA process, see below.

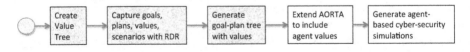

Fig. 5. Proposed process

Only the steps in grey background are considered in this paper and are briefly described as follows:

Step 1: The value tree would be provided by the individual or group whose values are being captured. Within an organization it is likely that there are multiple viewpoints and value-perspectives. It is likely that the CEO or Chief Information Officer would provide this direction to ensure decisions aligned with the goals of the organization. Using Schwartz's value and value categories, the individual would reflect on their abstract values and to add concrete values. The next section provides an example from a perspective provided in the literature.

Step 2: The capture of RDR rules, scenarios and associated values would probably be undertaken by the individual or group responsible for managing IT security for the organization. This could be a senior manager, or possibly the administrator for the corporate network and/or database. Then following the RDR revised knowledge acquisition process, explained below, each time one or more rules are added to cover a new case/situation, the user would also be prompted to select/add the name of the value and its degree of importance. An extension to RDR, known as C-MCRDR described below, also allows the cases to be incrementally captured so that new cybersecurity scenarios and the associated rules and values can be acquired as they arise.

Step 3: The cases, values and rules captured using C-MCRDR are used to generate a goal-plan tree. Following Sect. 2.3, (sub)goals are choices and in our approach are identified from the rule conclusion, whereas actions/plans are get methods for the rule conditions (attribute-value pairs) used to acquire the attribute values. The rule conclusion and conditions also correspond to the features of the case. Rules are also acquired that capture the weighting of relevant values. This information allows us to generate a goal-plan tree with values.

The focus in this paper is on the acquisition of the knowledge related to creation of an ethics in cybersecurity simulation, particularly on the use of RDR to assist KA. A modified RDR KA process is shown in Fig. 6. We see that additionally the domain expert/user is asked to review the value together with the case and conclusion and if they disagree with any of those they may select another conclusions and/or value or add a new conclusion, value and even a new case. If the conclusion has changed, then an additional step requires additional rule conditions to be added to distinguish this conclusion from the original. Multiple values can be added.

As a starting point for KA, RDR requires cases. Our approach supports two possibilities: 1) cases already exist, 2) cases need to be acquired. In some domains and some

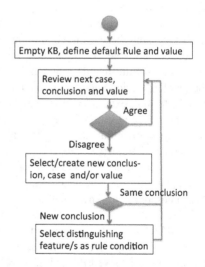

Fig. 6. RDR process including capture of values

problem types there will already be relevant cases. For example, an organization will store logfiles of the activities of their employees including logging in to various systems and networks and employee interactions involving the organizational email client. The IT department of the organization and/or the organization providing the email client may keep records on phishing attacks. A bank will keep track of network access and is likely to have records of network intrusion, attacks and other fraudulent activity.

To handle multiple, potentially conflicting, classifications of cases and to allow values to be added, we utilize an extension of MCRDR, C-MCRDR (also known as 7Cs [13]) which stands for Collaborative Configuration and Classification of a stream of incoming problem Cases via a set of Condition-Nodes linked to their Classes and associated Conclusions). C-MCRDR was created to support help desk applications where cases are worked up over a period of time and may involve multiple people. In C-MCRDR, a Classification is a class, grouping, category or a set of things sharing common properties, and may be labelled using text or hyperlinks, or it may remain unlabelled. This flexibility allows us to add the notion of values without the need for any extension to the method. C-MCRDR supports MCRDR-like case-driven KA and also supports top-down rule-driven KA [14]. This allows a value tree to be injected into the KB.

In [8] values, v, have a target value $T(v)$ that typically remain constant for the agent's life and also a Value State $S(v)$, to capture the current value in a given context. $T(v)$ would be captured as part of the KA process in a similar way that conclusions are added, while $S(v)$ would be captured at runtime within the cybersecurity simulation. We envisage that KA will be undertaken by domain experts, such as network or database administrators, who are globally distributed but who interact either due to belonging to the same organization or through belonging to special interest groups or online technical communities (e.g. Stackoverflow https://stackoverflow.com/.) The benefit of using these online communities to capture cases, conclusions and the values associated with the conclusions is that the landmark cases can be desensitized by removing specific details that would prevent disclosure while providing a valuable source of education and discussion among the relevant community. These cases can be used in formal, external and informal inhouse training courses teaching cybersecurity as a means to educate the next generation of software engineers, database administrators, network administrators, etc. This would assist in addressing the global shortage of adequately trained employees.

Value-based argumentation or deliberation tools, such as MOOD [ref] have been developed with the aim of allowing communities to agree on their shared values and their interpretation. The results of these processes can be incorporated in the KA process.

C-MCRDR has been designed to handle conflict management in previously seen cases, relationships between cases, rule nodes, conditions, classifications and conclusions. This is achieved through the use of live nodes and registered nodes, where the latter have been validated by a human user. Live nodes provide a public view, while registered nodes support private views that are valid/true for that user. A voting approach is used and, depending on the domain, certain users can be given different access rights, as described below, with different responsibilities for entering and accepting/authorising (some parts of) the knowledge base. Through the use of the live and registered status flags and the recording of the owner/author of all knowledge edits, conflicts between the public and private view/s are identified and brought to the users' attention for negotiation between the involved parties. More details are providing in [14, 15].

4   Example

For this paper we draw on discussion and examples provided in the Horizon 2020 report [4]. We consider the decision-making dilemma facing an organisation regarding mitigating potential cybersecurity attacks. As explained in [4], a number of viewpoints are possible, such as a shareholder viewpoint or a Corporate Social Responsibility (CSR) viewpoint, that lead to different values. Figure 7 shows the value tree using Schwartz's values from the shareholder perspective. We see that the shareholder perspective has a number of possible goals: seek for the organisation to be a market leader, seek for the organisation to be sustainable and responsible, minimise the risks and possibly shareholder liability and derive the most income from the shares. By traversing backwards from goals one can identify the concrete and abstract values associated with each goal. The value tree helps us to identify the nature of cases, conclusions and values to be captured for scenarios from this viewpoint.

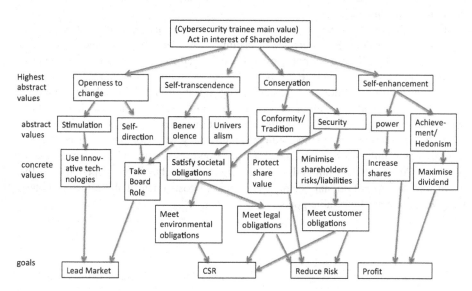

Fig. 7. Value tree for deciding how much to invest in cybersecurity attack prevention

A common dilemma faced by an organisation when a new threat is identified is how many resources (e.g. time, effort, people, technology, money) the organisation should invest in mitigating the threat. An organisation needs to consider whether it has the resources already, such as trained staff and appropriate technology, or whether it needs to draw on expertise outside the organisation. There are risks associated with this decision such as lack of sufficient internal expertise to handle the threat appropriately, leaking of knowledge and providing access to internal systems if involving an external party, extent of familiarity of a third party with the organisation and the risks associated with misunderstanding or miscommunication or unsuitable solutions. If choosing an external provider, there might be further considerations of the nature of the attack, method to be used, level of expertise, familiarity/relationship with the organisation and cost. For example, it might be better to use an external security organisation that you have worked with before successfully who are not very familiar with the attack to be mitigated than to work with a company who have expertise with that attack but with whom you have previous experience and face a steep learning curve to understand the systems in the organisation. The level of expertise of the provider and the method to be used is also likely to impact on the cost of the provider and the risk. The inhouse option is expected to have low cost, compared to an external provider. The organisation you have worked with before might have a medium cost, but as they lack experience in that threat the risk of a solution may be higher. In contrast, the company with high experience would be a low risk in finding a solution, but the cost of employing them may be high. Some of the variables from this scenario include: provider: {inhouse, externalNew, externalFamiliar}; method: {standard, new}, budget: {low, medium, high}. These variables can be used to create cases and identify values for each of the four goals {leader, CSR, risk, profit}. For simplicity the value of each of these goals and be {low, medium, high}.

For our example, we may have cases such as the five shown in Fig. 8. The C-MRDR mesh in Fig. 8 was created to capture Case 1 and its associated values. The mesh for the other 4 cases is not shown. Case 2 is added next to cover the situation where the method is new. Where the method is new an inhouse solution has higher risk. Case 3 considers the context where another organisation is considered for a standard task. Budget and their level of experience are also added as a case feature as it influences the decision. Case 4 captures the situation when Organisation A is not very experienced with the new method required. Familiarity with Company A is added as a condition as it influences the risk level. Case 5 captures the use of Company B as the provider. Company B is unfamiliar to the organization, but it has experience in the novel method. An additional negative is increased cost. Each of these factors potentially influence the four value goals.

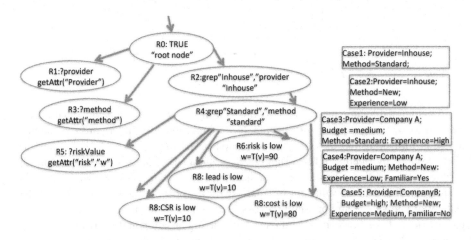

Fig. 8. C-MRCR Mesh snippet for Case 1. W = value of the weight which also represents the target value (T(v)).

Figure 8 only includes the rules for Case 1. Once we have the complete C-MCRDR mesh to cover the cases in Fig. 8, we can automatically generate the goal-plan tree in Fig. 9 (see step 3 description in Sect. 3). Then using the two-step process for making value-based decisions described in [8] we can first use the goal-plan-tree to generate a constraint problem and use a standard constraint solver to find the best course of action for that case. In the second step, the best course of action is implemented using extensions to AgentSpeak language as explained in [8] to support reasoning about values using the Jason language [16]. The captured values can be used to extend the AORTA framework so that the agent can reason about organisational norms while also making decisions using its own values.

For our scenario where an organization must decide what response to take to a particular cybersecurity attack there are three possible solutions involving different providers: inhouse, Company A, Company B. To make this decision the agent must gather some information. The information to be gathered may differ, as we found in the five cases captured using C-MCRDR. Figure 9 shows the goal tree for this decision and also includes the pre and post conditions. The pre and post conditions could also be explicitly captured via C-MCRDR, but we did not do so in our example. Value annotations V_i show the values for each option. To derive goal-plans-values, we elaborate the C-MCRDR tree for the five cases to derive goals = evaluateRisk, evaluateCSR, plan actions are getProvider, getMethod, getBudget, getExperience, getFamiliarity.

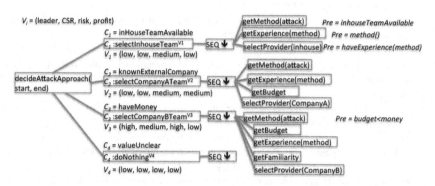

Fig. 9. Goal-Plan tree for deciding how much to invest in cybersecurity attack prevention

5 Discussion

Given the scarcity of cybersecurity scenarios and knowledge-based approaches to modelling value-based cybersecurity decision-making, we have proposed the use of RDR as a scalable and maintainable approach to capture the domain knowledge. Towards allowing a BDI agent to reason about the scenarios, we sketch how the RDR knowledge can be translated to a goal-plan tree. The approach used in [8] can be used to generate constraints and allow the agent to determine the best course of action. We propose to allow the agent to also consider organizational norms that the AORTA framework be modified to include the agents' values. This is left as future work.

6 Conclusion

Cybersecurity is a topic of major concern in many organisations. Support for decision making and training in cybersecurity are lacking. Given the vast scope and fast changing nature of dilemmas in this domain, we have sought to address the problem of how to acquire values and knowledge related to values that could be used by an agent. The approach proposed requires further elaboration and evaluation ideally via implementation and validation with domain experts and potential users of the scenarios. Based on this approach, a method can be developed to assist policy developers in stepwise design of complete policies.

References

1. Bowen, B.M., Devarajan, R., Stolfo, S.: Measuring the human factor of cyber security. In: 2011 IEEE International Conference on Technologies for Homeland Security (HST), pp. 230–235. IEEE (2011)
2. Hadlington, L.: The "Human Factor" in Cybersecurity. Psychological and Behavioral Examinations in Cyber Security, vol. 46 (2018)
3. Schwartz, S.H.: An overview of the Schwartz theory of basic values. Online Readings Psychol. Cult. **2**, 11 (2012)
4. Christen, M., Gordijn, B., Weber, K., van de Poel, I., Yaghmaei, E.: A Review of Value-Conflicts in Cybersecurity. ORBIT J. **1** (2017)
5. Jensen, A.S., Dignum, V., Villadsen, J.: A framework for organization-aware agents. Auton. Agent. Multi-Agent Syst. **31**, 387–422 (2017)
6. Rao, A.S., Georgeff, M.P.: Modeling rational agents within a BDI-architecture. In: Allen, J., Fikes, R., Sandewall, E. (eds.) Proceedings of the 2nd International Conference on Principles of Knowledge Representation and Reasoning, pp. 473–484. Morgan Kaufmann Publishers Inc. (1991)
7. Schwartz, S.H.: Are there universal aspects in the structure and contents of human values? J. Soc. Issues **50**, 19–45 (1994)
8. Cranefield, S., Winikoff, M., Dignum, V., Dignum, F.: No Pizza for You: Value-based Plan Selection in BDI Agents. In: Proceedings of the 26th International Joint Conference on Artificial Intelligence, pp. 178–184. AAAI Press (2017)
9. Richards, D.: Two decades of ripple down rules research. Knowl. Eng. Rev. **24**, 159–184 (2009)
10. Hofstede, G.: Dimensionalizing cultures: the Hofstede model in context. Online Readings Psychol. Cult. **2**, 8 (2011)
11. Spiekermann, S.: Ethical IT Innovation: A Value-Based System Design Approach. CRC Press (2015)
12. Poel, I.: Translating values into design requirements. In: Michelfelder, D.P., McCarthy, N., Goldberg, D.E. (eds.) Philosophy and Engineering: Reflections on Practice, Principles and Process. PET, vol. 15, pp. 253–266. Springer, Dordrecht (2013). https://doi.org/10.1007/978-94-007-7762-0_20
13. Vazey, M.M.: Case-driven collaborative classification (2007)
14. Vazey, M., Richards, D.: Achieving rapid knowledge acquisition in a high-volume call centre. In: Proceedings of the Pacific Knowledge Acquisition Workshop 2004, pp. 74–86. University of Tasmania Eprints Repository, Auckland (2004)

15. Richards, D.: A social software/Web 2.0 approach to collaborative knowledge engineering. Inf. Sci. **179**, 2515–2523 (2009)
16. Bordini, R.H., Hübner, J.F., Wooldridge, M.: Programming Multi-agent Systems in AgentSpeak Using Jason. Wiley (2007)

Towards Realtime Adaptation: Uncovering User Models from Experimental Data

Deborah Richards[1]($\boxtimes$), Ayse Aysin Bilgin[2], Hedieh Ranjbartabar[1], and Anupam Makhija[1]

[1] Department of Computing, Faculty of Science and Engineering,
Macquarie University, Sydney, Australia
deborah.richards@mq.edu.au, {hedieh.ranjbartabar,
anupam.makhija}@students.mq.edu.au
[2] Department of Statistics, Faculty of Science and Engineering,
Macquarie University, Sydney, Australia
ayse.bilgin@mq.edu.au

Abstract. Virtual Worlds and the non-player characters that inhabit them often lack knowledge about their users. Users are treated as sources of input or feedback. At best, systems respond to the user's behavioural data captured in logfiles. But there is no deep understanding of the player. Without this deep knowledge it is not possible for the computer to intelligently adapt. Relevant knowledge about the user will differ according to the application domain. Currently studies capture data such as biographical details, health status and history, psychological profiles, preferences and attitudes via questionnaires. This data can not be used in real time to influence the behaviour of the system. We suggest that data collected in past studies could be used to create user profiles and rules that can be used in real time for tailored interactions. We present two examples in this paper, one relating to an educational virtual world for science inquiry and the other involving the use of an Intelligent Virtual Agent to reduce study stress.

Keywords: Educational virtual worlds · Intelligent virtual agents
Knowledge acquisition · Data mining

1 Introduction

Adding intelligence to virtual worlds and the humanlike virtual agents that inhabit those worlds requires greater understanding of the player. Behavioural data and sensory input, even when responded to in real-time, is not enough – we need to know what motivates them, their state of mind, their medical history, their adherence history, whether they play computer games, their attitude to task, and so on. Building appropriate user profiles or models will depend on the role of the player. In the context of the serious games we are working with, the player could be a student, trainee or patient. To understand what features of the player influence how different players use our educational virtual worlds (EVWs) and interact with our intelligent virtual agents (IVAs) and to understand their experience and the relative benefits they derive from the

© Springer Nature Switzerland AG 2018
K. Yoshida and M. Lee (Eds.): PKAW 2018, LNAI 11016, pp. 46–60, 2018.
https://doi.org/10.1007/978-3-319-97289-3_4

experience, we have conducted a number of experiments. We have performed statistical analysis on the data to identify significant relationships. However, we have turned to data mining on the data collected from these studies to identify predictors and uncover rules that could be used by an intelligent tutoring system (such as an EVW) or IVA for reasoning about tailored interactions with users.

In the context of an IVA to reduce exam study stress, we ask questions such as:

What features of the player influence their preferences for a virtual character? What features of the player influence their attitude to the virtual character they meet and their responses to it? Which players need an empathic virtual character and benefit most from the interaction?

In the context of an EVW, we ask questions such as:

Can we predict their performance in a virtual world based on their individual features and/or behaviours while using the virtual world? Which players need educational scaffolding?

This paper does not seek to present answers to these questions, but to demonstrate how data can be used to answer such questions. In the next section we provide background to these two application domains to justify the data that was collected. In Sect. 3 we present the methodology. Results appear in Sect. 4, followed by discussion in Sect. 5, and conclusions and future directions in Sect. 6.

2 Identifying Data for Building User Models

Each application is likely to have data specific to a particular domain. For example, in our work that uses IVAs to encourage pediatric adherence to treatment advice [1], we capture data such as health history, treatments, health literacy, quality of life, adherence, health outcomes and patient experience.

One of the examples in this paper involves students using an EVW. EVWs use game technology to allow students to safely interact with objects and concepts being studied. Unlike a game, however, there are learning objectives to be achieved. We want learners to (re)construct their knowledge through exploration and experience, even if they might fail at times [2]. However, if students become stuck or disengaged it will be important to identify the learner's emotional state [3] and adapt the learning environment according to emotional needs [4–6]. Affect detection can involve human observation [7], self-reporting [8] or measuring physiological data [9, 10] including facial expressions [11, 12].

We are particularly interested in detecting engagement due to its important connection with learning [13–15]. Fredricks et al. [16] found a positive association between engagement and academic achievement, though other studies found weaker links [15, 17]. Appleton et al. [18] identifies four types of engagement: academic, behavioural, cognitive and psychological. To measure the latter two, Appleton et al. [18] introduced a survey instrument called Student Engagement Instrument (SEI). Cognitive engagement includes the constructs: Control and Relevance of School Work (CRSW), Future Aspirations and Goals (FAG), Extrinsic Motivation (EM) and Commitment to and Control over Learning (CCL). Psychological, or affective, engagement includes Teacher–Student Relationships (TSR), Peer Support for Learning (PSL), Family Support for Learning (FSL).

To measure behavioural engagement requires capture of data while the learner is engaged in a learning activity. In the context of an EVW, navigation paths are an indicator of performance [19] and level of engagement [20]. Navigation in virtual worlds may involve wayfinding, planning new routes and estimating directions. Comparison of navigation patterns within a virtual world has been investigated in the past to understand the influence of such behaviours on the user's performance. Significant differences have been observed in these patterns in terms of users' performance [21, 22]. Ruddle and Lessels [23] proposed three levels of performance metrics to evaluate the influence of navigation patterns on performance within a virtual world. These three levels involved metrics related to users' task performance, physical behaviour and cognitive behaviour. Task performance metrics might include measures such as task completion time, number of correct answers or distance travelled [23].

Learning and how it is achieved will also be influenced by individual factors such as demographics and personality [15, 24]. Numerous studies have examined the relationship between personality traits and performance [24–26]. For measuring personality, the most commonly used instrument is the Big Five Factor (BFF) model. The personality traits in this model are Openness, Conscientiousness, Extroversion, Agreeableness, Neuroticism sometimes also called OCEAN collectively [27, 28]. Openness trait relates to creativity while Neuroticism relates to emotional stability. Conscientious individuals have traits of being more organised, disciplined and hardworking. Agreeableness represents people with more sympathetic, thoughtful and cooperative nature. Extraverts are outgoing and friendly people who tend to form more social connections as compared to their introvert counterparts [29].

A related study done by Chamorro-Premuzic and Furnham [30] for university students emphasised that conscientiousness is a strong determinant of academic results. These ideas were further extended in another study in a school setting where they found that students with conscientious personality trait performed better in science subject [26]. Furnham and Monsen [26] also claimed that people with extravert personality are more interested in studies and perform better if they feel free and have more sense of freedom. Similar results have emerged from another study that points out that people with high level of openness and conscientiousness achieve success at university level [31]. In same direction, Eyong et al. [24] revealed that conscientiousness and agreeability traits are associated with the performance of secondary school students in positive manner and focused on promoting the need to encourage students to gain these behavioural traits for better academic achievements.

The psychological state of a student can affect their academic performance. A common way of measuring mental health in general, is the use of a psychometric test such as the Depression, Anxiety and Stress (DASS21) [32]. As mental health is a particular issue in the young adult age range, this data may also be relevant for understanding the performance of students and their responses to stressful event such as assessments and exams.

3 Methodology

We are conducting research using educational virtual worlds and intelligent virtual agents. The studies themselves and their results are not the focus of this paper. The goal of this paper is to look at how data mining can be used for extracting models and rules from the data collected in our experimental studies in order to make our educational virtual worlds and intelligent agents more intelligent and adaptive to users' characteristics, behaviours and needs. The generic process in Fig. 1 shows how data captured within a study (including pre, during and post data) is analysed and the combined data used for machine learning.

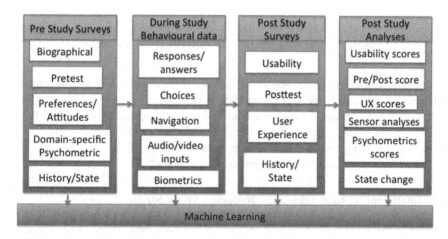

Fig. 1. Generic data capture and analyses process

The classification method, C 5.0 Decision Tree (Hastie et al., 2009, Han and Kamber, 2006) is used to identify the relative predictive value of each of the variables on the categories of outcome variable(s). A classification model was considered successful when it consistently returned high predictive accuracy rates across all categories of outcome variable(s). The model predictions were compared to random predictions (class specific lifts) to decide which model was best for future predictions. In addition, class specific accuracies were compared to identify the best classification model. All classification models were developed using IBM SPSS Modeler 18.0 (IBM, 2016).

C5.0 decision tree(s) were grown for the Study Stress study dataset by setting the pruning severity to 75% (or lower depending on the outcome variable) and minimum records for each leaf to 5 (or 3 depending on the outcome variable). Since we did not have a test data set, and the study data set was small, we used ten-fold cross-validation as a safe guard for overfitting the model(s) for the study data set.

In Fig. 2 we see a sample IBM SPSS Modeler stream using the EVW SPSS.sav file as the source. The input fields and target to be predicted are selected in the "Type" node. The example shows two different targets: Score 1 and Score 2 from the EVW study. The Auto Classifier modelling node was first used to identify that C5.0 was the

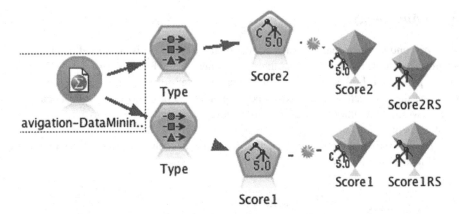

Fig. 2. Method for generating rules using IBM SPSS Modeler.

best method to use. The diamonds show the models developed for each target using C5.0 and the associated rule sets (Score1RS, Score2RS).

In both studies survey data was captured online via the Qualtrics research software (Qualtrics.com). Demographic data collected included gender, age, cultural background, degree being studied and computer game activity. Below we describe data specific to each experiment and the approach to data analysis.

3.1 Intelligent Virtual Agents for Reducing Study Stress

Before interaction with the IVA, we captured attitude to study (using 5pt Likert scale from strongly dislike to strongly like), their preference for the character's age (peer-age, younger, doesn't matter), gender (male, female, doesn't matter) and ethnic appearance (same, different, doesn't matter) and if it should look like them (yes, no, doesn't matter). To identify if the humans' personality influenced their preferences and responses to the IVAs we captured the BFF via the 50-item International Personality Item Pool (IPIP), 10 items measuring each of the five personality scales [33]. We used DASS21 questionnaire to capture the psychological emotional state of the user [32].

To measure the efficacy of our IVA to change study stress levels, we also captured level of study stress (on a scale of 0–10) before the experiment and after each interaction with the IVA. Zero means "extremely good and relaxed" and 10 means "extremely bad and stressed". To measure the human's response to the IVA, we asked whether they liked the character (or control document) and used a rapport questionnaire to measure their perceived sense of rapport with the character or document (control).

Next, participants received tips to reduce study stress through interaction with one of our IVAs while the control group read a pdf on study tips. The users' keyboard and mouse interaction data with the IVA were captured into a separate MySQL database.

3.2   Educational Virtual World for Science Inquiry

We created the Omosa Virtual World, an EVW to help students gain biology knowledge and science inquiry skills. Participants experienced two different scenarios in our EVW. After the first scenario they answered a 10 question quiz about what they had learnt. Then they interacted with a second scenario and answered another quiz with 10 questions.

The navigation data of the participants was logged during interaction by collecting their position every second. Each of these data points were combined to form a full navigation path for each participant. The navigation data was recorded at regular intervals so that not only the paths were captured, but also the time how long the participant took to complete the task. The conversations with virtual characters were logged as well, to allow further insight into the behaviour of the participants.

After each scenario, we asked participants how much on a scale of 1 to 5 they found the experience enjoyable, with 1 being not enjoyable, and 5 enjoyable. Participants were also asked to elaborate why they thought the experience was enjoyable or not. The participants were also asked for usefulness of the hints, if they belonged to a group where hints were given for second scenario. These strategies provide us with the possibility for better comprehension of the behaviour of participants and more sophisticated interpretation of data during analysis.

After using the EVW, we asked participants to complete a 10-item Personality Traits Questionnaire [28]. To measure cognitive and affective engagement we used the Student Engagement Instrument (SEI) [18]. SEI was designed for use on students in middle school, so we modified it for university students. For example, we replaced the word "school" with university. SEI includes 32 items, nineteen items to measure affective engagement and fourteen to measure cognitive engagement. Participants responded to statements by rating their agreement on Likert scale of 1–4 with 1 indicating that they never felt or acted that way, and 4 indicating that they certainly acted that way.

The student's behavioural engagement was captured via their navigational interactions within the world and their academic engagement was captured via the number of correct answers to the quizzes given after using the EVW. The virtual world was deliberately laid out in such a way that participants may not always know what to do, or where to go next.

3.3   Data Processing and Analysis

The data was cleaned to remove responses that were mostly incomplete. Personality, DASS21, Quiz and SEI scores were computed and added as new variables. Categories were created for many factors to allow crosstabulations and chi-square tests from continuous data. For example we used the number of correct answers to the quiz taken after using the EVW to categorise participants into groups to represent their performance (Low = 1 to 4 correct; Medium = 5 to 6 correct; High = 7 or more correct).

Descriptive statistics, crosstabulations and chi-square tests were performed using the reporting and data analysis tools in Qualtrics. A series of t-tests and ANOVAs were performed in IBM SPSS statistical package (IBM, 2016a). With these tools we found many significant results, but we were not able to develop a coherent model of the combination of variables. For this we turned to IBM SPSS Modeler (IBM, 2016b).

4   Results

The Study Stress experiment involved 239 participants, (165 females, 72 males, 2 other) aged between 17–46 (mean age = 20.23, SD = 3.64). The largest cultural group represented South-East Asian (22.18%), followed by Northern-Western European (17.15%), and North African and Middle Eastern (12.55%). Oceania (which includes Australia) formed a small proportion of participants (4.18%). Most of the participants were enrolled in a Psychology degree (74.48%) and less than half of the participants regularly played computer games (43.1%). The majority of participants' were neutral about study (48.95%) while 23.01% and 4.18% of students' chose like and love, and 19.67% and 3.35% of students' chose dislike and hate, respectively.

In the EVW study we had 115 participants, comprised of 78 females and 37 males, ages ranging from 17–33 (mean 19.80 s.d. 2.84). The largest cultural groups identified as "Northern Western European (17.74%), "Oceania" (which includes Australia) (16.13%) and "South East Asia" (16.13%) and 19.35% did not identify with any cultural group. Only 43.48% of participants played computer games, for an average of 7.35 h per week, ranging from 1 h to 50 h (s.d 8.24).

For space and scope reasons, we do not present further biographical or survey results, such as DASS21, SEI, OCEAN/BFF. We also do not present behavioural logfile data or quiz, rapport or study stress scores that were collected at multiple timepoints. These results will be published in other articles. Below we present selected results from data mining as examples of knowledge discovery made possible by data mining.

4.1   Reducing Study Stress Results

We considered many target variables to model. Along the lines of the questions posed in the introduction, target variables considered were those that measured preferences (IVA age, gender, ethnicity, similarity), the relationship (IVA rapport and liking), and benefit (change in stress level, change in quiz score). As input variables we include demographics (age, gender, culture, study field, game playing, attitude to study), DASS21 and personality.

Below we show the evaluation of the C 5.0 decision tree produced by using IBM SPSS Modeler for responses to the question "Would you prefer a virtual character to be: younger than you, same age as you, or older than you? Table 1 shows counts and percentages for each response as observed (in the rows) and predicted (in the columns). A chi-square test shows a highly significant result for this model (χ^2 = 30.601, df = 2, p < 0.001). The highlighted cells give us class specific accuracies for each outcome category. The model is unable to make any prediction for "younger than you" category therefore the accuracy is zero while 99.4% of "same as you" and 18.4% of "older than you" were correctly predicted by the C5.0 decision tree. The total accuracy of the model is 78.9%. The lift of a specific outcome category calculated dividing the proportion of correct predictions for a specific category by the proportion of this category in the data set. For example for the "same age as you" lift is calculated as following (178/227)/(179/237) = 1.04. The idea here is to compare how many the model gets correct out of the all predictions for that category against how many we would expect to be correct if we had randomly sampled from the data set. The values higher than 1 is desirable while values close to 1 indicate the model is as good as random allocations. In Table 1, we observed that "older than you" lift is 4.35 which means the model is 4.35 times better than random allocation.

Table 1. Count and percentages for preferences for age of virtual character

Would you prefer a virtual character to be?		Younger than you	Same age as you	Older than you	Total	Accuracy %	Lift
Younger than you	Count	**0**	9	0	9	0.0	
	Row %	0	100	0	100		
	Column %	0	3.96	0	3.80		
	Total %	0	3.80	0	3.80		
Same age as you	Count	0	178	1	179	99.4	1.04
	Row %	0	**99.44**	0.56	100		
	Column %	0	78.41	10	75.53		
	Total %	0	75.11	0.42	75.53		
Older than you	Count	0	40	9	49	18.4	4.35
	Row %	0	81.63	**18.4**	100		
	Column %	0	17.62	90	20.68		
	Total %	0	16.88	3.80	20.68		
Total	Count	0	227	10	237	78.9	
	Row %	0	95.78	4.22	100		
	Column %	0	100	100	100		
	Total %	0	95.78	4.22	100		

In Fig. 3 we see that the most important predictor variable is participant gender (57.95%), followed by Culture (15.51%), Neuroticism (14%), Extroversion (8.04%) and less so, attitude to study (3.17%) and whether they liked the character (1.32%). Figure 4 shows the decision tree produced by C5.0.

As another example, we show the Modeler output for responses to the question "Do you prefer the character to look like you?" with responses yes, no, doesn't matter. A chi-square test shows a highly significant result for this model ($\chi^2 = 49.816$, df = 4, p < 0.001). In Fig. 5 we see that the most important predictor variable is whether the participant plays computer games (yes, no; 31.93%), followed by culture (20.47%),

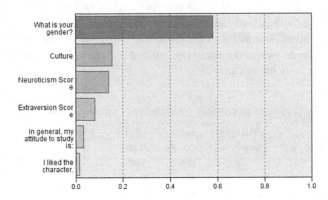

Fig. 3. Predictor importance participants' preference for character age

What is your gender? in [Female] [Mode: 2] (164)
 Culture in [Oceania Southern-Eastern European North_east Asian Southern and Central Asian People of the Americas
   Sub-Saharan African I don't identify with any cultural group] [Mode: 2] => Same age as you (62.532; 0.864)
 Culture in [Northern-Western European] [Mode: 2] (34.216)
  In general, my attitude to study is: in [Hate Dislike] [Mode: 2] => Same age as you (5.46; 0.702)
  In general, my attitude to study is: in [Neutral Like Love] [Mode: 2] (28.755)
   I liked the character. in [Strongly Disagree Disagree Neutral] [Mode: 3] (7.756)
    Extraversion Score <= 33 [Mode: 2] => Same age as you (4.374; 0.771)
    Extraversion Score > 33 [Mode: 3] => Older than you (3.383; 0.887)
   I liked the character. in [Agree Strongly Agree] [Mode: 2] => Same age as you (20.999; 0.847)
 Culture in [North African and Middle Eastern] [Mode: 2] (25.957)
  Neuroticism Score <= 19 [Mode: 3] => Older than you (3.449; 0.87)
  Neuroticism Score > 19 [Mode: 2] => Same age as you (22.507; 0.839)
 Culture in [South-East Asian] [Mode: 2] (41.295)
  In general, my attitude to study is: in [Hate Dislike Neutral Like] [Mode: 2] => Same age as you (37.043; 0.764)
  In general, my attitude to study is: in [Love] [Mode: 3] => Older than you (4.252; 0.765)
What is your gender? in [Male Don't identify with either] [Mode: 2] => Same age as you (73; 0.699)

Fig. 4. Decision tree for participant's preference for the age of the virtual character

attitude to study (20.13%), grade they expect to achieve (9.22%), gender (8.49%), neuroticism (personality) score (5.25%), and the experimental group (4.51%) (whether they received empathic or neutral IVA first or were in control group who received the document with tips).

4.2   Educational Virtual World Results

In the EVW context we were interested to learn the rules for participant performance (quiz scores, time taken) and benefit of receiving hints versus not receiving hints (change in quiz score). In Fig. 6 we see the decision tree for the class Score 1.

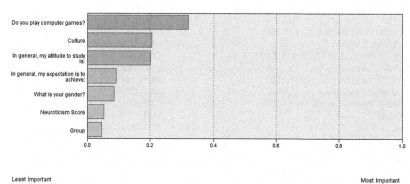

Fig. 5. Predictor importance for target variable "Do you want the character to look like you?"

```
Chars <= 5 [ Mode: 1 ] (82.434)
        PlayGames = 2.000 [ Mode: 1 ] => 1.0 (52.717; 0.829)
        PlayGames = 1.000 [ Mode: 1 ] (29.717)
                Backtrack1 = TRUE [ Mode: 2 ] => 2.0 (12.297; 0.593)
                Backtrack1 = FALSE [ Mode: 1 ] => 1.0 (17.42; 0.861)
Chars > 5 [ Mode: 3 ] (32.566)
        CRSW <= 2.860 [ Mode: 3 ] (29.283)
                TSR <= 3.330 [ Mode: 3 ] (24.283)
                        TSR <= 2.780 [ Mode: 2 ] (9.283)
                                TSR <= 2.670 [ Mode: 1 ] => 1.0 (2; 0.5)
                                TSR > 2.670 [ Mode: 2 ] (7.283)
                                        CRSW <= 2.290[Mode:3]=>3.0 (2; 1.0)
                                        CRSW>2.290 [M:2]=>2.0(5.283; 0.811)
                        TSR > 2.780 [ Mode: 3 ] => 3.0 (15; 0.867)
                TSR > 3.330 [ Mode: 2 ] => 2.0 (5; 0.8)
        CRSW > 2.860 [ Mode: 1 ] => 1.0 (3.283; 0.695)
```

Fig. 6. Decision tree for Quiz 1 Score (1 = low; 2 = medium, 3 = high)

The C5.0 model for Score 1 uncovers that the number of characters talked to (51%) is the most important predictor variable, followed by whether participants backtracked (22%), whether they play computer games (16%), teacher-student relationship (TRS) affective engagement score (11%) and Control and Relevance of School Work (CRSW) cognitive engagement score (only 1%). The rules for the decision tree in Fig. 6 are shown in Fig. 7.

```
Rules for 1 - contains 5 rule(s)              and Backtrack1 = TRUE
  Rule 1 for  1.0 (49.717; 0.839)               then 2.000
    if Chars <= 5                             Rule 3 for  2.0 (2; 1.0)
    and Teleport1 = FALSE                       if Chars <= 5
    and PlayGames = 2.000                       and Teleport1 = TRUE
    then 1.000                                  and Age > 19
  Rule 2 for  1.0 (17; 0.882)                   then 2.000
    if Chars <= 5                             Rule 4 for  2.0 (5.283; 0.811)
    and Teleport1 = FALSE                       if Chars > 5
    and PlayGames = 1.000                       and CSRW <= 2.860
    and Backtrack1 = FALSE                      and TSR <= 3.330
    then 1.000                                  and TSR <= 2.780
  Rule 3 for  1.0 (4; 1.0)                       and TSR > 2.670
    if Chars <= 5                               and CSRW > 2.290
    and Teleport1 = TRUE                        then 2.000
    and Age <= 19                             Rule 5 for  2.0 (5; 0.8)
    then 1.000                                  if Chars > 5
  Rule 4 for  1.0 (2; 0.5)                       and CSRW <= 2.860
    if Chars > 5                                and TSR > 3.330
    and CSRW <= 2.860                           then 2.000
    and TSR <= 3.330                        Rules for 3 - contains 2 rule(s)
    and TSR <= 2.780                          Rule 1 for  3.0 (2; 1.0)
    and TSR <= 2.670                            if Chars > 5
    then 1.000                                  and CSRW <= 2.860
  Rule 5 for  1.0 (3.283; 0.695)                and TSR <= 3.330
    if Chars > 5                                and TSR <= 2.780
    and CSRW > 2.860                            and TSR > 2.670
    then 1.000                                  and CSRW <= 2.290
Rules for 2 - contains 5 rule(s)                then 3.000
  Rule 1 for  2.0 (0.717; 1.0)              Rule 2 for  3.0 (15; 0.867)
    if Chars <= 5                               if Chars > 5
    and Teleport1 =                             and CSRW <= 2.860
    then 2.000                                  and TSR <= 3.330
  Rule 2 for  2.0 (9; 0.667)                    and TSR > 2.780
    if Chars <= 5                               then 3.000
    and Teleport1 = FALSE                   Default: 1
    and PlayGames = 1.000
```

Fig. 7. Rules for Quiz 1 Score

5   Discussion

Interpretation of the results produced will be essential to determine the value of the modeling process. Our purpose is to find rules that we can implement in our virtual world and IVA architectures that allow them to provide tailored experiences for the user according to the users' need, preference and characteristics. The example provided using the data from the Study Stress experiment shows us that feature of the player that influence their preference for a virtual character include participants' gender, culture, neuroticism and extraversion personality traits and attitude to study and whether they play computer games. To determine the importance of these factors we need to look at how many people are covered by a particular rule and the rule accuracy. From the decision tree in Fig. 4, gender is the only predictor for males or "don't identify with either", with 69.9% of the 73 participants (male or don't identify with either) having chosen "same age as you" for virtual character age. For the 164 females, other factors were also taken into account. The ramification of this model is that if we are able to provide different virtual characters, we should provide males with a character the same age. Regarding looks, gender and ethnicity we need to review the other outputs for each of these target variables. We note that gender was also a predictor (8.49%) for whether the character should look similar to the participant, but the decision tree (not shown) identified 11 participants (male or don't identify with either) with 74% accuracy who said similar appearance did not matter.

From the example provided using the participants' score in Quiz 1 after using the EVW, we found that factors relating to the task (number of characters the participant speaks to) and participant behaviour in the EVW (whether they backtrack or not) influence the score as well individual factors such as whether they play computer games and to a smaller extent the participants' affective and cognitive engagement. By running models for all of the target variables, we can build up a picture to identify why tips would be helpful (e.g. don't backtrack) and for which participants. This can make learning more tailored and efficient as we do not want to provide all possible tips to players or interfere with the learning process of those students who are able to work out good game-play and learning strategies without our hints.

The significance of certain variables on the performance, preferences, behaviours and experiences of users as uncovered using IBM SPSS Modeler is not surprising. Our studies were designed by drawing on the literature and thus the data we captured which form the input to our models is likely to be relevant. Data mining has helped us understand when certain data is relevant and can also help us better design future studies by removing data that is not predictive or useful for understanding our participants and the results of our studies. Cronbach and Snow [34] suggest that it is often not possible to take the results of one study and apply them to another study when the subject matter or the set of participants are (significantly) different, and that such analyses may have to be done for each study. For this reason, the models we seek to build are based on features of the participants and we only intend to apply the model to new participants also with those features. In the current work we have not considered including features of the scenario or subject matter. Given that individuals have different needs and/or responses in different contexts, we may also need to include not only the participant features but also features of the scenario in which the model was captured to identify if the model will apply when the subject matter changes.

6 Conclusions and Future Directions

In this paper we demonstrated how we mine the survey and behavioural data (data captured via interaction log files) to gain knowledge about our users so that we might create user models to be used in real-time to provide tailored responses. We are rerunning some of our studies to ensure that the models we develop to represent the varying needs and preferences of user cohorts are not overfitted to current datasets. We are also exploring how these models could be uncovered in real-time and progressively adapted as more participant data becomes available and even over a period of interaction with the user.

With the rules uncovered through data mining, we will update our EVW to include hints and run an experiment with all, tailored and no (control) hints to see the effect on performance (quiz score and time taken). For the IVA work, we intend to update the Fatima emotion appraisal and cognitive agent architecture [35] with a user model containing relevant user features, add rules and modify existing components accordingly, to allow the IVA to adapt to the needs of their user to deliver greater therapeutic benefit. Discovering these rules and important user features has been made possible through data mining.

Acknowledgements. This work was partially supported by the Australian Research Council Discovery Grant DP150102144 "Agent-based virtual learning environments for understanding science". Thanks to participants in all of the studies.

References

1. Richards, D., Caldwell, P.: Improving health outcomes sooner rather than later via an interactive website & virtual specialist. IEEE J. Biomed. Health Inf. (2017)
2. Kapur, M.: Productive failure. Cogn. Instr. **26**, 379–424 (2008)
3. Shen, L., Wang, M., Shen, R.: Affective e-learning: Using" emotional" data to improve learning in pervasive learning environment. J. Educ. Technol. Soc. **12**, 176 (2009)
4. Arguel, A., Lane, R.: Fostering Deep Understanding in Geography by Inducing and Managing Confusion: An Online Learning Approach (2015)
5. Ranjbartabar, H., Richards, D.: Student designed virtual teacher feedback. In: Proceedings of the 9th International Conference on Computer and Automation Engineering, pp. 26–30. ACM (2017)
6. Arguel, A., Lockyer, L., Lipp, O.V., Lodge, J.M., Kennedy, G.: Inside out: detecting learners' confusion to improve interactive digital learning environments. J. Educ. Comput. Res. **55**, 526–551 (2017)
7. Woolf, B., Burleson, W., Arroyo, I., Dragon, T., Cooper, D., Picard, R.: Affect-aware tutors: recognising and responding to student affect. Int. J. Learn. Technol. **4**, 129–164 (2009)
8. Sabourin, J., Mott, B., Lester, J.: Computational models of affect and empathy for pedagogical virtual agents. In: Standards in emotion modeling, Lorentz Center International Center for workshops in the Sciences (2011)
9. Prendinger, H., Ishizuka, M.: The empathic companion: a character-based interface that addresses users'affective states. Appl. Artif. Intell. **19**, 267–285 (2005)
10. D'mello, S.K., Graesser, A.: Multimodal semi-automated affect detection from conversational cues, gross body language, and facial features. User Model. User-Adap. Inter. **20**, 147–187 (2010)
11. Ammar, M.B., Neji, M., Alimi, A.M., Gouardères, G.: The affective tutoring system. Expert Syst. Appl. **37**, 3013–3023 (2010)
12. Duo, S., Song, L.X.: An e-learning system based on affective computing. Phys. Procedia **24**, 1893–1898 (2012)
13. Hassaskhah, J., Khanzadeh, A.A., Mohamad Zade, S.: The relationship between internal forms of engagement (Cognitive-Affective) and academic success across years of study. Issues Lang. Teach. **1**, 251–272 (2013)

14. Hoff, J., Lopus, J.S.: Does student engagement affect student achievement in high school economics classes. In: The Annual Meetings of the Allied Social Science Association, Philadelphia, PA (2014)
15. Carini, R.M., Kuh, G.D., Klein, S.P.: Student engagement and student learning: testing the linkages. Res. High. Educ. **47**, 1–32 (2006)
16. Fredricks, J.A., Blumenfeld, P.C., Paris, A.H.: School engagement: Potential of the concept, state of the evidence. Rev. Educ. Res. **74**, 59–109 (2004)
17. Ewell, P.: An Analysis of Relationships Between NSSE and Selected Student Learning Outcomes Measures for Seniors Attending Public Institutions in South Dakota. National Center for Higher Education Management Systems, Boulder, CO (2002)
18. Appleton, J.J., Christenson, S.L., Kim, D., Reschly, A.L.: Measuring cognitive and psychological engagement: validation of the student engagement instrument. J. Sch. Psychol. **44**, 427–445 (2006)
19. Hanna, N., Richards, D., Jacobson, Michael J.: Academic performance in a 3D virtual learning environment: different learning types vs. different class types. In: Kim, Y.S., Kang, B.H., Richards, D. (eds.) PKAW 2014. LNCS (LNAI), vol. 8863, pp. 1–15. Springer, Cham (2014). https://doi.org/10.1007/978-3-319-13332-4_1
20. Hanna, N., Richards, D., Hitchens, M., Jacobson, M.J.: Towards quantifying player's involvement in 3D games based-on player types. In: Proceedings of the 2014 Conference on Interactive Entertainment, pp. 1–10. ACM, Newcastle (2014)
21. Darken, R.P., Sibert, J.L.: Navigating large virtual spaces. Int. J. Hum. Comput. Interact. **8**, 49–71 (1996)
22. Sas, C., O'Hare, G., Reilly, R.: A performance analysis of movement patterns. In: Bubak, M., van Albada, G.D., Sloot, Peter M.A., Dongarra, J. (eds.) ICCS 2004. LNCS, vol. 3038, pp. 954–961. Springer, Heidelberg (2004). https://doi.org/10.1007/978-3-540-24688-6_122
23. Ruddle, R.A., Lessels, S.: Three levels of metric for evaluating wayfinding. Presence Teleoperators Virtual Environ. **15**, 637–654 (2006)
24. Eyong, E.I., David, B.E., Umoh, A.J.: The influence of personality trait on the academic performance of secondary school students in Cross River State, Nigeria. IOSR J. Humanit. Soc. Sci. **19**, 12–19 (2014)
25. Fosse, T.H., Buch, R., Säfvenbom, R., Martinussen, M.: The impact of personality and self-efficacy on academic and military performance: The mediating role of self-efficacy. J. Mil. Stud. **6**, 47–65 (2015)
26. Furnham, A., Monsen, J.: Personality traits and intelligence predict academic school grades. Learn. Individ. Differ. **19**, 28–33 (2009)
27. Goldberg, L.R.: An alternative" description of personality": the big-five factor structure. J. Pers. Soc. Psychol. **59**, 1216 (1990)
28. Gosling, S.D., Rentfrow, P.J., Swann, W.B.: A very brief measure of the Big-Five personality domains. J. Res. Pers. **37**, 504–528 (2003)
29. McCrae, R.R., Costa Jr., P.T.: A five-factor theory of personality. Handb. Pers. Theory Res. **2**, 139–153 (1999)
30. Chamorro-Premuzic, T., Furnham, A.: Personality, intelligence and approaches to learning as predictors of academic performance. Pers. Individ. Diff. **44**, 1596–1603 (2008)
31. Hazrati-Viari, A., Rad, A.T., Torabi, S.S.: The effect of personality traits on academic performance: The mediating role of academic motivation. Procedia Soc. Behav. Sci. **32**, 367–371 (2012)
32. Henry, J.D., Crawford, J.R.: The short-form version of the Depression Anxiety Stress Scales (DASS-21): construct validity and normative data in a large non-clinical sample. Br. J. Clin. Psychol. **44**, 227–239 (2005)

33. Goldberg, L.R.: A broad-bandwidth, public domain, personality inventory measuring the lower-level facets of several five-factor models. Pers. Psychol. Eur. **7**, 7–28 (1999)
34. Cronbach, L.J., Snow, R.E.: Aptitudes and Instructional Methods: A Handbook for Research on Interactions. Irvington, Oxford (1977)
35. Dias, J., Mascarenhas, S., Paiva, A.: FAtiMA Modular: Towards an Agent Architecture with a Generic Appraisal Framework. In: Bosse, T., Broekens, J., Dias, J., van der Zwaan, J. (eds.) Emotion Modeling. LNCS (LNAI), vol. 8750, pp. 44–56. Springer, Cham (2014). https://doi.org/10.1007/978-3-319-12973-0_3

Supporting Relevance Feedback with Concept Learning for Semantic Information Retrieval in Large OWL Knowledge Base

Liu Yuan[✉]

College of Computer Science, Shaanxi Normal University, Xi'an, China
yuanliu@snnu.edu.cn

Abstract. Relevance feedback in information retrieval is a popular way to learn the user's intents. We investigate the feasibility and methodology of applying the concept learning in OWL knowledge base to deal with feedback in interactive information retrieval system. The feedback from the initial search results is considered as examples, and then the inductive concept learning technique is employed to generate a concept that describes the user's requirement. We deal with the performance of concept learning by reducing the scale of the problem; a clustering based OWL knowledge base partitioning method is proposed to divide the knowledge base into several small-scale ones with acceptable recall and precision. An interactive healthcare information retrieval prototype is developed for evaluation; the results of the user study and precision-recall graph show the efficiency of the methods proposed.

Keywords: OWL knowledge base · Information retrieval
Relevance feedback · Concept learning

1 Introduction

Recent behavioral studies show a large portion of user information-seeking activities are exploratory and characterized by the complex and evolving nature of user information needs [1, 2]. Relevance feedback is a direct way to know user's information needs [3]. In OWL-based semantic retrieval environment [4], the resources are described by concepts and relations that defined in ontologies, so the resources can be considered as instances of concepts. If we apply relevance feedback to this environment, the relevant or non-relevant resources (examples) user indicated can be considered as instances of a concept that can describe user's need, the goal of search is to find all the instances of the concept in the data collection.

Learning from small datasets and the use of background knowledge to construct explanations are a distinctive feature of Inductive Learning (IL) [22], and Inductive Logic Programming (ILP) is an example of IL that is the intersection of machine learning and logic programming [18, 19]. For an example-guided semantic retrieval, the users can specify their search intent using positive and negative examples making ILP a possible fit. OWL coincides with the description logic, a fragment of first order logic, so in theory, the methods from ILP are applicable to concept learning in OWL

© Springer Nature Switzerland AG 2018
K. Yoshida and M. Lee (Eds.): PKAW 2018, LNAI 11016, pp. 61–75, 2018.
https://doi.org/10.1007/978-3-319-97289-3_5

knowledge base [5, 6, 26], and it is possible to apply this technique to example-guided semantic retrieval system for learning the hidden concept that examples belong to. Application of concept learning method is a complicated procedure, there are two major problems we should deal with when implementing the procedure. One is to represent the resources fully and correctly with semantic information defined in the background knowledge base according to the requirement of concept learning. Another problem is the performance of the concept learning algorithm, the efficiency of the reasoning service that determines the performance of the concept learning may not meet the user's expectation. For large knowledge base, it is necessary to preprocess the knowledge for improving the performance of learning procedure. In this paper, we will investigate these problems and focus on how to make OWL knowledge base fit the concept learning procedure well by utilizing its structure features.

The rest of the paper is organized as follows: In Sect. 2, we overview the related work. The fundamental techniques about our work are explained in Sect. 3. In Sect. 4, we focus on solving performance problem by reducing the scale of concept learning problem, the detailed implementation algorithms are presented. In Sect. 5, we analyze the experiment data and results intensively. Finally, the conclusions and future work are drawn in Sect. 6.

2   Related Work

Our work intersects multiple research areas. We introduce the related work from the following relevant research aspects.

In the area of user interface and semantic retrieval, a number of user interfaces have been designed to facilitate construction of queries for better search results [9, 10, 13]. These interfaces can help users to describe their intention more precisely and explicitly, but require users to use a term that is compatible with the data schema and background knowledge [11, 15]. An interactive interface is always provided to enhance the semantic search. For example, a system can allow users to start a search by entering a keyword query, then guides users through an incremental construction process, which quickly leads to the desired semantic query [10]. Users are assumed to have basic domain knowledge but do not need specific details of the ontology, or proficiency in a query language. There exists work that combining results categorization and personalized information retrieval to introduce a novel personalized concept-based search mechanism, for example, users can use keyword and results are organized into concepts using special conceptual vocabularies [12]. All the research mentioned above have proved the feasibility and efficiency of combining semantic information and interactive search mode.

The two questions about background knowledge for concept learning that arise naturally are what should be coded and how should it be coded. ILP-based methods have proved the effective way to implement relevance feedback in information retrieval system [14, 15]. An earlier research work [14] has proposed a prototype system enabling a user to search for scientific papers by positive example. That search procedure can be naturally converted into an ILP problem without requiring any background knowledge for users. However, we cannot apply ILP methods to the OWL knowledge base directly because of the disparity in knowledge representation. Some research have focused on the mining Semantic Web data using ILP-based idea [16], and there are some work about concept learning algorithm for the description logics based on refinement operator originated from ILP [5–8], the research results showed that in principle it is feasible to implement interactive search by using concept learning algorithm on OWL knowledge base, but there is still lack of referential work focusing on how to improve the performance of concept learning based application by utilizing the features of knowledge base from the perspective of application.

Dealing with large OWL knowledge base is a challenge for researchers in both big data and Semantic Web areas. Parallel processing is the most popular and effective way to deal with big data [24], but this philosophy does not works for OWL datasets directly because of the complicated relationship between data. There is research focusing on dividing a large OWL dataset into small-scale sets without losing important semantic information [25], but how to solve this problem for semantic information retrieval have not been fully investigated.

3 Concept Learning Problem in OWL Knowledge Base

3.1 Concept Learning Problem

Basically, concept learning for OWL knowledge base can be considered as concept learning for description logics knowledge base, as defined in Definition 1 [5].

Definition 1 (Concept Learning in Description Logics, CLDL). Let a concept name *Target*, a knowledge base K (not containing *Target*), and the sets E_+ and E_- with elements of the form $Target(a)$ ($a \in I$, I is the set of individuals of K) be given. The learning problem is to find a concept C such that Target does not occur in C and for $K' = K \cup \{Target \equiv C\}$ we have

(1) $K' \models E_+$ (completeness)
(2) $K' \not\models E_-$ (consistency)

Such a concept C is called correct.

In Definition 1, $\models$ denotes the classical logical consequence. The background knowledge is composed of *TBox* and *ABox*. The *TBox* describes the terminology by relating concepts and roles; the *ABox* contains assertions about objects, it relates objects to concepts and other objects via roles. We ignore the negative examples for making the problem simpler.

Example 1. Consider a simple task of learning a concept *relevant* from a given query result, defined on document *d*. The value of *relevant(d)* is true if *d* is relevant to a query. Figure 1 shows an example to illustrate this task. The training examples consist of four positive examples and four negative examples. Each positive example is an instance of the concept *relevant* that is known to be true, while each underlined negative example indicates a document are not relevant to the given query. The background knowledge includes concepts, relationships defined in *TBox* and instances defined in *ABox*, which are represented by the following concepts: HeartFailure,

Cardiomegaly, DrugTherapy, Physiopathology, Pathology and DrugEffect and relation hasTopic, respectively. The output of a CLDL algorithm could be the concept description: $HeartFailure \sqcap \exists hasTopic.$ ($DrugTherapy \sqcup PhysioPathology$). The constructed concept simply states that if a document d is an instance of both atomic concept $HeartFailure$ and constructed concept $\exists hasTopic.$ ($DrugTherapy \sqcup PhysioPathology$), then the document d satisfies the current query.

Background Knowledge

HeartFailure, Cardiomegaly, DrugTherapy, PhysioPathology, Pathology, DrugEffect, hasTopic.

HeartFailure(d_0), HeartFailure(d_1), HeartFailure(d_2), HeartFailure(d_3), HeartFailure(d_5), HeartFailure(d_8), Cardiomegaly(d_7),

Training Set

$relevant(d_0)$ $relevant(d_5)$
$relevant(d_1)$ $relevant(d_6)$
$relevant(d_2)$ $relevant(d_7)$
$relevant(d_3)$ $relevant(d_8)$

hasTopic(d_0, DrugTherapy), hasTopic(d_1, DrugTherapy), hasTopic(d_2, PhysioPathology), hasTopic(d_3, Pathology), hasTopic(d_6, DrugTherapy), hasTopic(d_7, DrugTherapy), hasTopic(d_5, DrugEffect).

Fig. 1. An example of concept learning in description logics

3.2   The General Procedure of CLDL Based Interactive Search

The CLDL based interactive search is a four-step procedure: (1) Performing normal search on the knowledge base; (2) According to the hitting list, user inspects several top-ranked resources and marks those with relevance/positive(E_+) or irrelevance/negative(E_-)(optional); (3) Based on E_+, E_- (optional) and knowledge base K, system automatically learns the target concept of user's intention; (4) Performing the new search, that is finding the instances of target concept in the knowledge base.

This procedure is repeated as long as there are new relevant documents near the top of the list or the user gives up. It is intuitive that the user's positive/negative feedback from the previously search result corresponds to the E_+/E_- in the concept learning task. In terms of knowledge base K, it is not directly perceived by an information retrieval system, so we should construct a knowledge base to prepare for CLDL.

Basically, constructing a knowledge base for CLDL includes two sub-tasks: defining *TBox* and generating *ABox*. For *TBox*, we need related domain knowledge to determine the concepts and roles that can describe the area of the problem; For *ABox*, we need to create the instances of concepts and roles defined in *TBox*. Now we illustrate how to construct a knowledge base with an example in healthcare domain.

Example 2. Constructing a knowledge base about healthcare information.

(1) **Define *TBox*:** The domain knowledge we used for defining terminologies come from MeSH thesaurus. The MeSH thesaurus is a controlled vocabulary with RDF format produced by NLM (National Library of Medicine).

(2) **Generating *ABox*:** In our examples, the objects in the *ABox* are documents in healthcare domain from which we search. So generating *ABox* is to annotate the documents with terms defined in *TBox* [23]. An example of annotating document d with four mesh terms is displayed in Fig. 2: document d is annotated with four

single MeSH terms "Enlarged Heart", "Heart Enlargement", "Cardiac Hypertrophy" and "Heart Hypertrophy". Because MeSH provides several popular and easy-understandable terms for each professional medical concept, we use this information to annotate the documents, the procedure is described in Algorithm 1.

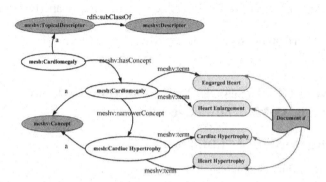

Fig. 2. Annotating the Resources with the Terminologies in the Knowledge base. The vocabularies defined in MeSH RDF Schema ("meshv:TopicDescriptor", "meshv:Descriptor", "meshv:Concept", "meshv:hasConcept", etc.) are used to define professional medical concepts ("mesh: Cardiomegaly", "mesh: Cardiac Hypertrophy", "Enlarged Heart", etc.) and relations between concepts that corresponding to *TBox* terminologies, the documents are annotated by the terms appeared in *TBox*.

Algorithm 1: Document Annotation by Terminologies in *TBox*
Input: Documents set *R*, Ternimonogies in *TBox*
Output: Annotation results *S*
1. for each document *d* in *R*
2.   *Td* = findMeSHVocabulariesByTerms(*d*); //return all MeSH vocabularies in *d*, both popular terms and professional notations
3.   for each *t* in *Td*
4.   if there exist a term defined in TBox semantically equivalent to *t*
5.   annotating *d* with *t*;
6. $S = S \cup \{(d, ht)\}$;
8.   endfor
9. endfor
10. return *S*;

(3) The knowledge base *K* with respect to Fig. 2 has the following form:
Concepts and Roles in *TBox* as background knowledge: Cardiomegaly; Cardiac Hypertrophy; hasConcept(Cardiomegaly, Cardiomegaly); narrowConcept (Cardiomegaly, Cardiac Hypertrophy); hasTerm(Cardiomegaly, Enlarged Heart); hasTerm(Cardiomegaly, Heart Enlargement); hasTerm(Cardiac Hypertrophy, Cardiac Hypertrophy); hasTerm(Cardiac Hypertrophy, Heart Hypertrophy).

Assertions in *ABox*: Cardiomegaly(*d*); Cardiac Hypertrophy (*d*).

4 Improving CLDL Search Performance by Reducing the Scale of OWL Knowledge Base

4.1 Reducing the Scale of CLDL Problem

The CLDL algorithm determines the performance of an example-guided semantic retrieval system, one of the key factors that affect CLDL is the reasoning capability of description logics. While scalability of description logics reasoning system has improved substantially in the recent years, OWL and related description logics are inherently difficult to reason with, as their worst-case complexity classes are ExpTime or worse. Therefore, reducing the scale of CLDL is a simple and practical way to improve the performance. Since the number of examples is quite small, the scale of CLDL will be reduced if we only use a subset of the knowledge base that meets the need of a given query. That means we need to partition the knowledge base into small-scale subsets, only relevant subsets will be used when performing CLDL based search. We give the following definition of knowledge base partition:

Definition 2 (Knowledge Base Partition). Given a description logics based logics knowledge base $K = <TBox, ABox>$, a partition P_K of the knowledge base K is a set of subset of K with $P_K = \{K_1, K_2, \ldots, K_n\}$, $K_i = <TBox_i, ABox_i>$, $TBox_i \subseteq TBox$, $ABox_i$, $i = 1, 2, \ldots, n$.

Definition 3 (Acceptable Knowledge Base Selection). Let a description logics based knowledge base K, a subset of K AP_K and a query Q be given. $Ans(K, Q)$ represent the correct answer set of Q on K. AP_K is an acceptable selection for Q if $\frac{|Ans(AP_k,Q)|}{|Ans(K,Q)|} \geq \delta$, δ is the acceptance threshold given by the user.

It is obvious that the acceptable knowledge base selection is a tradeoff between accuracy and executing time. The goal of selection is to find a subset of a knowledge base with respect to a query acceptable. However, we cannot determine whether a selection is acceptable or not before processing a query, so we need to find a way to approximate an acceptable selection for query processing.

4.2   Clustering Based Partition by Analyzing the Structure of Knowledge Base

Within data analysis, cluster analysis has always been very popular in grouping data. Analyzing the structural features of OWL knowledge base will help us to build clusters. When using RDF data model to picture the OWL knowledge base, we find that clusters can be characterized by the structure of the RDF triples [21].

4.2.1   Analyzing the Cluster Structure of OWL Knowledge Base

The basic representation of RDF data model is RDF triple with the form (s, p, o), where (s, p, o) stands for $(subject, predicate, object)$. The clusters on the RDF triples have the following types:

(1)   The triples in the cluster have the same $subject$ s: Let $CType_1$ be a cluster of this type if $\forall(s_j, p_j, o_j), (s_k, p_k, o_k) \in CType_1$, then $s_j = s_k$ holds.

(2)   The triples in the cluster have the same $predicate$-$object$ pattern (p, o): Let $CType_2$ be a cluster of this type if $\forall(s_j, p_j, o_j), (s_k, p_k, o_k) \in CType_2, j \neq k, s_i \in S, s_j \in S$, then $p_j = p_k, o_j = o_k$ hold. Where S is the set of subjects in the cluster.

(3)   The $predicate$-$object$ of the triples in the clusters come from the same pattern set $\{(p_1, o_1), \ldots, (p_f, o_f)\}$: Let $CType_3$ be a cluster of this type, if $\forall(s_j, p_j, o_j) \in CType_3$, then there exists a triple $(s_j, p_k, o_k) \in CType_3$, $(p_k, o_k) \in \{(p_1, o_1), \ldots, (p_f, o_f)\}, k \neq j, s_j \in S$. Where S is the set of subjects in the cluster.

We can cluster the OWL knowledge base using these characters of clusters type. In general, the size of $ABox$(the number of instances) is much larger than that of $TBox$(the number of concepts), so basically, the knowledge base clustering is the partition of $ABox$. The instances in $ABox$ correspond to the subject s in RDF triples directly, we can cluster the $ABox$ by the features of these instances. For identifying the instances within the same cluster, we need compare the value of $predicate$-$object$ pair (p, o). For the convenience of computation, we re-represent the instances in the ABox in a new form. An instance can be described by a feature set fs, every feature in fs has an associated value set. fs has the form $fs = \{f_1, \ldots, f_n\}$, $\forall f_i \in rfs_k$ there is a $Vf_i = \{vf_1, \ldots, vf_k\}$.

Example 3.   Representing the instance in Example 2 in the form of fs.
In Example 2, there is a resource d in ABox, let the fs of d be fsd = {hasDescriptor, hasTerm}, then Vf_d^1 = {Cardiomegaly, Cardiac Hypertrophy}, Vf_d^2 = {Enlarged Heart, Heart Enlargement, Cardiac Hypertrophy, Heart Hypertrophy}. If there is another resources d', let the fs of d' be fsd' = {hasDescriptor, hasTerm}, then $Vf_{d'}^1$ = {Cardiomegaly, Blood Vessels}, $Vf_{d'}^2$ = {Enlarged Heart, Heart Hypertrophy, blood}. We can compare the d and d' based on this expression.
The similarity between the value of feature sim_v is defined as:

$$sim\_v(vf_1, vf_2) = \begin{cases} 1 & if \ vf_1 = vf_2 \\ 0 & else \end{cases}$$

Hence the similarity between the features sim_f is defined as:

$$sim\_f(f_1,f_2) = \frac{\sum_{s=1}^{k} \max(sim\_v(v_s,v_t))\forall t \in [1,h]}{k},$$

Where $v_s \in Vf_1, v_t \in Vf_2, k = |Vf_1|, h = |Vf_2|$.
The sim_f is asymmetry, that means if $k \neq h$ then $sim\_f(f_1,f_2) \neq sim\_f(f_2,f_1)$.

4.2.2   Partitioning OWL Knowledge Base by Clustering
According to the discussion above, we can find a model that characterizes a cluster for each cluster type using feature set expression fs of the instance, and then cluster the knowledge base by these cluster models. An acceptable selection with respect to a query will be acquired by combining several clusters.

Algorithm 2: Finding Cluster Models
Input: The set R of $ABox$ instances with form fs;
Output: Cluster models $CT1$, $CT2$ and $CT3$ for $CType_1$, $CType_2$ and $CType_3$ respectively.
1. $CT1= \emptyset, CT2 = \emptyset, CT3 = \emptyset$;
2. for each instance s in $ABox$
3. $CT1= CT1 \cup \{(s)\}$; // generating mode of $CType_1$
4. for each feature f in R $CType_1$
5.   for each value vf of f
6. $CT2 = CT2 \cup \{(f_i, vf_i)\}$; // generating mode of $CType_2$
7. Sub = all subsets of $CT2$;
8. for each se in Sub
9. $CT3 = CT3 \cup se$; // generating mode of $CType_3$
10. Return $CT1$, $CT2$, $CT3$;

The goal of clustering knowledge base is to find all possible clusters whatever the type, the procedure of clustering is described as follows.

Algorithm 3. Clustering the Knowledge Base
Input: Knowledge base K, $CT1$, $CT2$ and $CT3$ generated by Algorithm 2;
Output: Preliminary Knowledge Base Partition P_K
1. $P_K = \emptyset$, $CT = CT1 \cup CT2 \cup CT3$;
2. for each model m in CT {
3. Sub_A = all assertions meet m in $ABox$; // generating ABox of the partition with respect to cluster model m
4. Sub_T = all terminologies related with Sub_A; // generating TBox with respect to Sub_A
5. $K_m =\{ <Sub_T, Sub_A>\}$;
6. $P_K = P_K \cup K_m$;
7. Return P_K;

The P_K generated from Algorithm 3 shows a distinguishing feature: $P_K = \{K_1, K_2, \ldots K_n\}$, for partitions $K_i \in P_K$, $K_j \in P_K$, $i,j = 1, 2, \ldots, n$; it is reasonable that $K_i \cap K_j \neq \varnothing$. The overlapping between the partitions in P_K is intended to keep the semantic integrity over the sub-knowledge base. For processing a query, we can select one or more partitions from P_K and combine them as a subset of the knowledge base that is complete enough to deal with the query. If we select these partitions appropriately, we can get an approximately acceptable selection. Therefore, a method of finding an approximately acceptable selection is proposed by combing the preliminary clusters in P_K, as described in Algorithm 4.

Algorithm 4: Finding an Acceptable Selection for a query
Input: Clustering based Partition P_K, a query q;
Output: Approximately Acceptable Knowledge Base Partition AP_K for q.
1. $AP_K = \varnothing$;
2. for each K_i in P_K
3.   if there exist a terminology in K_i corresponding to keywords(or user's feed back) of q;
4. $AP_K = AP_K \cup K_i$;
5. Eliminating the duplicated facts and rules in AP_K;
6. rerurn AP_K;

The essential task of constructing an acceptable knowledge base partition is to find the relevance between preliminary clusters and the relation between query keywords and every preliminary cluster. An easy and understandable technique to measure the relevance between entities is to use similarity measure methods in data management field, both in a syntactical way and in a semantical way. Because this task is highly domain dependant, the relationships between terminologies in a certain domain can provide valuable information to judge the relevance. In this paper, we make fully use of the MeSH tree structure and the properties of MeSH vocabularies, such as "seeAlso", "preferredTerm", "preferredConcept" etc., to identify the relevance between different entities.

5 Experiment

5.1 Dataset and Evaluation Criteria

We collected a total of 21,000 Web pages from WebMD.com, a well-known health information service website, and selected 1,275 web documents covering 3 topics of ADHD, allergy, and anxiety-panic as evaluation pages, and design 2 query tasks for each topic as evaluation task. The correct answer of each query is identified manually and used for calculating recall and precision. 600 concepts and 2350 relations (roles) derived from MeSH vocabularies and MeSH tree structure are defined in the complete *TBox*, we only consider the MeSH resources that are relevant to the pages we collect (at least 27,336 concepts will be defined if we used the whole MeSH descriptions). 21,000 instances of concepts and 86,000 instances of relations are asserted in the complete *ABox*.

For each topic, we divided the document set into two collections, one which is used for the initial query and relevance judgments, and the second that is then used for comparative evaluation. The performance of both initial query and feedback based query are compared to the second collection. The reason for that experiment plan is to pursue fairness in the evaluation. According to the example-guided search procedure described above, the concept learning based query processing is based on the feedback from the initial query result. So it is obvious that we can start with an initial query and to compute a precision-recall graph. Following one round of feedback from the user, we perform the concept learning based query processing and again compute a precision-recall graph. It is straightforward to compare the two performances for evaluation, but it is unreasonable. Because the gains of the feedback based query are partly due to the fact that known relevant documents (judged by the user) are now ranked higher. Fairness demands that we should only evaluate with respect to documents not seen by the user.

We also evaluated the effectiveness through user studies, particularly by doing a time-based comparison: since we have answers to every query task, we make the comparison of how many relevant documents a user found in a certain amount of time by different search strategy. Such notions of user utility are much fairer to real system usage. We evaluate the effect of the method proposed by combining precision-recall graph and user studies together. The recall and precision are defined as: $recall = |A \cap A_r|/|A_r|$, $precison = |A \cap A_r|/|A|$, where A_r is the set of correct answers identified manually and A is the set of answers returned by the CLDL based search. All tests were run on a 2.4 GHz Core i7 core machine with 8 GB RAM.

5.2   Experiment Results

5.2.1   The CLDL Based Search on the Complete Knowledge Base

We developed a prototype of healthcare information retrieval system to prove the efficiency of the CLDL based search strategy. Figure 3 is the search interface returned by the system to a query of "insect allergy": users can input the query keywords into the text box just as keyword based normal search, or input query by clicking the professional terms shown on the right of each result or the general descriptions on the left of the results. The distinguishing feature is the checkbox in front of the each search

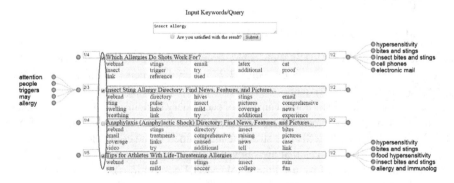

Fig. 3. An Interactive Search Interface for Healthcare Information Retrieval

result, users can mark the checkbox for expressing their positive evaluation of the current search. The results displayed in this interface are only from one of the two collections with respect to the topic user concerned when formulating a query.

We recruited 5 adults for the user study, each participant was required to find some information about the search tasks devised for the experiment according to the content of the documents in the data collection. The participants completed the search task by both normal search interface and feedback supported search interface. We recorded the query user formulated and relevant documents user marked on the interface for performing the concept learning based search in the other collection of the concerned topic user. All the participants agree that the search interface can support feedback in a simple but effective way and can help a lot to formulate a query that expresses their intentions correctly. We use the CELOE algorithm in [8] to implement the concept learning task, the main algorithm parameters are set with default values. Reasoning request on the knowledge base is accomplished by approximate reasoner considering the efficiency.

Table 1(a) illustrates the average performance of normal search and concept learning based search respectively according to the participant's experimental record.

Table 1(a). Average Performance of CLDL Based Search (Recall and Precision are represented in fraction form, N-S stands for Normal Search, CLDL-S stands for CLDL based Search)

Topics		Recall		Precision		Running Time(s)	
		N-S	CLDL-S	N-S	CLDL-S	N-S	CLDL-S
ADHD	Q_1^1	16/16	16/16	16/23	16/18	0.42	2.9
	Q_1^2	6/7	6/7	6/10	6/6	0.39	2.7
Allergy	Q_2^1	25/27	27/27	25/36	27/29	0.47	3.6
	Q_2^2	11/12	11/12	11/16	11/11	0.44	3.3
Anxiety	Q_3^1	19/19	19/19	19/42	19/23	0.43	3.1
	Q_3^2	4/5	5/5	4/10	5/5	0.56	3.8

For each topic, there are 2 tasks to be completed by querying the system. Since we only have relatively small documents set for evaluation, the recall and the precision are expressed in fraction form instead of percentage form, the numerator and the denominator represent the number of documents corresponding to the recall and precision formula mentioned above respectively. It is comprehensible that the improvement of recall and precision on CLDL based search mainly attribute to the precise expression of the user's intention in an interactive way. That improvement shows the results returned by this interactive search are more pertinent to user's intention.

Normal search means search without considering relevance feedback, the system processes the user's query only using query keywords, but the documents retrieved are still annotated semantically with terminologies in the knowledge base. The semantic annotation of the documents is also the reason for the good normal search results. Running time for normal search is the response time from the query submitted to the

Table 1(b). Average Number of Documents Participants Found in 3 min (N-S stands for Normal Search, CLDL-S stands for CLDL based Search).

Topic		Number of Documents found	
		N-S	CLDL-S
ADHD	Q_1^1	7 of 16	11 of 16
	Q_1^2	4 of 7	4 of 7
Allergy	Q_2^1	14 of 27	19 of 27
	Q_2^2	4 of 12	6 of 12
Anxiety	Q_3^1	15 of 19	16 of 19
	Q_3^2	3 of 5	3 of 5

search results returned; time for CLDL based search we concerned includes the time consumed for both concept learning and query processing. It is a conspicuous disadvantage of the interactive search system.

From a human-computer interaction perspective, the CLDL based search can help users to find the information they need faster. Table 1(b) is the comparison of the average number of the relevant documents participants found in 3 min using normal search strategy and concept learning based search strategy respectively.

For implementing the example-guided search, it is necessary to preprocess the knowledge base and resources to be retrieved. The preprocessing tasks include making the format of background knowledge understandable for CLDL algorithm, annotating the resources with terminologies and grouping the knowledge base by cluster analysis etc. Although the preprocessing is time-consuming and complicated, all the tasks would be completed before the query processing, so the search performance will not be influenced.

5.2.2 The Search Based on Partitioned Knowledge Base

The performance showed in Table 1(a), 1(b) worked on the complete knowledge base, in this section we discuss how much the knowledge base partition affects the system performance. We generated preliminary clusters of knowledge base firstly, and then the approximately acceptable knowledge base partition is generated dynamically when performing interactive query processing. We found 13,204 preliminary clusters on the whole 21,000 web pages by using Algorithm 2 and Algorithm 3, which means not all concepts are used for our documents set. The same tasks settings with Table 1(a), 1(b) are used to test the concept learning based interactive search on partitioned knowledgebase. We selected several preliminary clusters and constructed an acceptable subset of the knowledge base for each query task by using Algorithm 4.

In order to prove the effect of knowledge base partition further, we also execute the same task by normal search on the partitioned knowledge base; the acceptable knowledge selection is constructed only by the user's input keywords. The performance of the normal search and the interactive search on the partitioned knowledge

Table 2(a). The Performance of Normal Search on Partitioned Knowledge Base (Recall and Precision are represented in fraction form, C-KB stands for Complete KB, P-KB stands for Partitioned KB)

Topic		Recall		Precision		Time(s)	
		C-KB	P-KB	C-KB	P-KB	C-KB	P-KB
ADHD	Q_1^1	16/16	16/16	16/23	16/23	0.42	0.29(1272)
	Q_1^2	6/7	6/7	6/10	6/10	0.39	0.18(853)
Allergy	Q_2^1	25/27	25/27	25/36	25/36	0.47	0.24(2018)
	Q_2^2	11/12	11/12	11/16	11/16	0.44	0.19(972)
Anxiety	Q_3^1	19/19	19/19	19/42	19/42	0.43	0.22(1054)
	Q_3^2	4/5	4/5	4/10	4/10	0.56	0.17(933)

Table 2(b). The Performance of CLDL based Search on Partitioned Knowledge Base (Recall and Precision are represented in fraction form, C-KB stands for Complete KB, P-KB stands for Partitioned KB)

Topic		Recall		Precision		Time(s)	
		C-KB	P-KB	C-KB	P-KB	C-KB	P-KB
ADHD	Q_1^1	16/16	16/16	16/18	16/18	2.9	1.2(292)
	Q_1^2	6/7	6/7	6/6	6/6	2.7	0.8(188)
Allergy	Q_2^1	27/27	27/27	27/29	27/29	3.6	1.4(415)
	Q_2^2	11/12	11/12	11/11	11/11	3.3	1.2(372)
Anxiety	Q_3^1	19/19	19/19	19/23	19/23	3.1	1.2(274)
	Q_3^2	5/5	5/5	5/5	5/5	3.8	1.6(219)

base is presented in Table 2(a) and 2(b) respectively, and the number in parentheses indicates how many preliminary clusters are used for the current query task.

It is obvious that the partitioned knowledge base can shorten the running time with a remaining accuracy for the tasks we designed, either for normal search or CLDL based search. This improvement can be ascribed to the reduced knowledgebase scale, the acceptable partition generated using Algorithm 4 contains the enough resources information for every single query. In interactive information retrieval environment, the scale of the partitioned knowledge base is much smaller than normal search because we further select more relevant information from the knowledge base according to user's feedback. We noticed that the reduced running time is not proportional to the scale of the knowledge base, for example, running on the 188 of 13,204 preliminary clusters consumed nearly 1/3 running time on the whole knowledge base. The reason for that is the relevance computation in Algorithm 4 spent much time, so further optimization is required.

6 Conclusion and Further Work

This paper demonstrates the feasibility of interactive information retrieval system based on concept learning in OWL knowledge base and proposes a knowledge base partition method based on RDF data clustering technique to improve the performance of the interactive search. The RDF/OWL data representation for knowledge base has proved effective both in understanding the search intention and improving the search results. Having all the necessary elements in place, we incorporated the methodologies from different research area into an interactive search mechanism that will be made available to users for healthcare information search. The main advantage of the method proposed is that we can explain and describe user's search intent explicitly and in a formal way, so the IR system can benefit from this feature. Clustering based dataset dividing also fully used the characters of the OWL data, so it can reduce the scale of IR without losing recall and precision. However, the disadvantages are also obvious. Compared with classical IR strategies, concept learning based IR cost more resources and time because of the efficiency of the description logics reasoner. We still can not apply our work to web-scale applications directly.

Further work will focus on the scalability and the optimization of the concepts learning based information retrieval system. First of all, we need to construct a complete and objective benchmark for evaluating the CLDL based information retrieval. Although a knowledge base partition method is proposed to deal with performance problem in this paper, dynamically selecting the sub-knowledge base still need to be optimized further. The examples that come from the user's feedback are very limited in quantity, although inductive learning has the ability to deal with small data, the more facts about the examples will be helpful in finding the accurate concepts. But how to find additional facts appropriately without affecting the performance is still under the investigation, it is what we will focus on in the future.

Acknowledgement. Supported by the Fundamental Research Funds for the Central Universities (No. GK201503066)

References

1. Ruotsalo, T., et al.: Interactive intent modeling: Information discovery beyond search. Commun. ACM **58**(1), 86–92 (2015)
2. Marchionini, G.: Exploratory search: from finding to understanding. Commun. ACM **49**(4), 41–46 (2006)
3. Manning, C.D., Raghavan, P., Schtze, H.: Relevance feedback and query expansion. Introduction to Information Retrieval. Cambridge University Press, New York (2008)
4. AlObaidi, M., Mahmood, K., Sabra, S.: Semantic enrichment for local search engine using linked open data. In: Proceedings of the 25th International Conference Companion on World Wide Web. International World Wide Web Conferences Steering Committee, pp. 631–634 (2016)
5. Bühmann, L., Lehmann, J., Westphal, P.: DL-Learner—A framework for inductive learning on the Semantic Web. Web Semant. Sci. Serv. Agents World Wide Web **39**, 15–24 (2016)

6. Lehmann, J., Hitzler, P.: Concept learning in description logics using refinement operators. Mach. Learn. **78**(1–2), 203–250 (2010)
7. Fanizzi, N., d'Amato, C., Esposito, F.: DL-FOIL concept learning in description logics. In: Železný, F., Lavrač, N. (eds.) ILP 2008. LNCS (LNAI), vol. 5194, pp. 107–121. Springer, Heidelberg (2008). https://doi.org/10.1007/978-3-540-85928-4_12
8. Lehmann, J., Auer, S., Bühmann, L., et al.: Class expression learning for ontology engineering. Web Semant. Sci. Serv. Agents World Wide Web **9**(1), 71–81 (2011)
9. Zenz, G., Zhou, X., Minack, E., Siberski, W., Nejdl, W.: Interactive query construction for keyword search on the semantic web. In: De Virgilio, R., Guerra, F., Velegrakis, Y. (eds.) Semantic Search over the Web. Data-Centric Systems and Applications, pp. 109–130. Springer, Heidelberg (2012). https://doi.org/10.1007/978-3-642-25008-8_5
10. Zenz, G., Zhou, X., Minack, E., et al.: From keywords to semantic queries—incremental query construction on the Semantic Web. Web Semant. Sci. Serv. Agents World Wide Web **7**(3), 166–176 (2009)
11. Bobed, C., Esteban, G., Mena, E.: Enabling keyword search on Linked Data repositories: an ontology-based approach. Int. J. Knowl. Based Intell. Eng. Syst. **17**(1), 67–77 (2013)
12. Sah, M., Wade, V.: Personalized concept-based search on the Linked Open Data. Web Semant. Sci. Serv. Agents World Wide Web, **36**, 32–57 (2016)
13. Caruccio, L., Deufemia, V., Polese, G.: Understanding user intent on the web through interaction mining. J. Vis. Lang. Comput. **31**, 230–236 (2015)
14. Dimec, J., Dzeroski, S., Todorovski, L., et al.: WWW search engine for Slovenian and English medical documents. Stud. Health Technol. Inf. **68**, 547–552 (1998)
15. Loggie, W.T.H.: Using inductive logic programming to assist in the retrieval of relevant information from an electronic library system. Notes of the Workshop on Data Mining, Decision Support, Meta Learning and ILP held at The Fourth European Conference on Principles of Data Mining and Knowledge Discovery, Lyon, Ftance (2000)
16. d'Amato, C., Fanizzi, N., Fazzinga, B., Gottlob, G., Lukasiewicz, T.: Semantic web search and inductive reasoning. In: Bobillo, F., et al. (eds.) UniDL/URSW 2008-2010. LNCS (LNAI), vol. 7123, pp. 237–261. Springer, Heidelberg (2013). https://doi.org/10.1007/978-3-642-35975-0_13
17. Fazzinga, B., Gianforme, G., Gottlob, G., Lukasiewicz, T.: Semantic Web search based on ontological conjunctive queries. J. Web Sem. **9**(4), 453–473 (2011)
18. Lavrac, N., Dzeroski, S.: Inductive Logic Programming. WLP, pp. 146–160 (1994)
19. Džeroski, S.: Relational Data Mining. Springer, US (2009)
20. Medical Subject Headings. https://www.nlm.nih.gov/mesh/
21. Giannini, S.: RDF data clustering. In: Abramowicz, W. (ed.) BIS 2013. LNBIP, vol. 160, pp. 220–231. Springer, Heidelberg (2013). https://doi.org/10.1007/978-3-642-41687-3_21
22. Gulwani, S., Hernandez-Orallo, J., Kitzelmann, E., Muggleton, S.H., Schmid, U., Zorn, B.: Inductive programming meets the real world. Commun. ACM **58**(11), 90–99 (2015)
23. Berlanga, R., Nebot, V., Pérez, M.: Tailored semantic annotation for semantic search. Web Semant. Sci. Serv. Agents World Wide Web **30**, 69–81 (2015)
24. Kaminsky, A.: BIG CPU, BIG DATA: Solving the World's Toughest Computational Problems with Parallel Computing (2016)
25. Jagvaral, B., et al.: Large-scale incremental OWL/RDFS reasoning over fuzzy RDF data. In: 2017 IEEE International Conference on Big Data and Smart Computing (BigComp). IEEE (2017)
26. Lisi, F.A.: A formal characterization of concept learning in description logics. In: 25th International Workshop on Description Logics (2012)

Combining Concept Learning and Probabilistic Information Retrieval Model to Understand User's Searching Intent in OWL Knowledge Base

Liu Yuan[(⊠)]

College of Computer Science, Shaanxi Normal University, Xi'an, China
yuanliu@snnu.edu.cn

Abstract. Understanding and describing user's searching intent in exploratory information retrieval is a key issue for improving the relevance of search results. Employing concept learning method and probabilistic information retrieval model, this paper proposes an exploratory information retrieval strategy that can explain user's search intent in a formal way. User's relevance feedback from the initial search results are considered as examples and the user's searching intent is described as concepts learned from the knowledge base and examples. Uncertain inference with respect to the concept learned in knowledge base is used to implement probabilistic information retrieval. By constructing a probabilistic OWL knowledge base, this paper develops a healthcare interactive information retrieval prototype to evaluate the method proposed. The experiment results prove the advantages of using concept learning in exploratory semantic retrieval.

Keywords: Concept learning · Semantic Web
Exploratory information retrieval · Probabilistic model · Semantic retrieval

1 Introduction

The research in the past few decades have shown that information retrieval(IR) can benefit much from the Semantic Web technology in understanding the meaning of resources [1, 2], but it is still not easy to describe the user's search intent explicitly. Relevance feedback is a direct way to learn the user's searching intent [3, 4]. In semantic retrieval environment, the resources are usually annotated with concepts implied in OWL ontologies, so the resources can be considered as instances of concepts, and the reasoning capability of ontology could be utilized when finding instances of a given concept [5, 6]. If we adopt relevance feedback information to the semantic retrieval environment, the relevant or non-relevant resources user indicated in the feedback can serve as positive and negative examples for supervised learning, and the learning result can be considered as concepts for describing user's intent. Relevance feedback provides many challenges for supervised learning. For example, users have little patience with examining non-relevant documents, so learning is typically done

© Springer Nature Switzerland AG 2018
K. Yoshida and M. Lee (Eds.): PKAW 2018, LNAI 11016, pp. 76–89, 2018.
https://doi.org/10.1007/978-3-319-97289-3_6

with the small, nonrandom documents set that results from the system's best attempt to retrieve relevant documents.

Learning from small datasets and the use of background knowledge to construct explanations are distinctive features of Inductive Learning(IL) [7], and Inductive Logic Programming(ILP) is an example of IL that is the intersection of machine learning and logic programming [8]. The users can specify their searching intent using positive and negative examples making ILP a possible fit. OWL coincides with the description logic, a fragment of first order logic, so the methods from ILP are applicable to concept learning in OWL knowledge base [9, 10], and it is possible to apply this technique to semantic retrieval for learning the hidden concept that can describe user's intent. However, logic by itself cannot fully model IR. In determining the instances of the concepts learned, the success or failure of an implication relating the instances and concepts is not enough. It is necessary to take account the uncertainty inherent in such an implication. Probabilistic information retrieval is a IR model to score queries by a criteria motivated by probability theory, and it usually is adopted to deal with relevance feedback in query processing.

The research objective and highlights of this paper are listed as follows: (1) Concept learning can find a concept to describe user's intent from very limited feedback information and express the searching intent in a formal way, that make user's intent explicit. (2) The uncertain inference related to the concepts learned makes probabilistic information retrieval model applicable to the OWL-based knowledge base.

2 Related Work

Our work intersects multiple research areas; we introduce the related work from the following relevant research aspects:

Ontology Based Information Retrieval: Making use of the knowledge representation and reasoning capability of ontology, researchers have applied ontology to IR successfully [11]. Most of the ontology based IR focused on how to exploit the concepts and relations between concepts defined in ontology, usually the query terms are mapped to concepts and the relations between concepts, such as synonymy, hyponymy, and instantiation, meronymy and similarity. These concepts and relations are used to expand both queries and document indexing entries [12].

Interactive Search Interface and User's Intent Modeling: User's relevance feedback comes from an effective interactive search interface, it is also the fundamental of user's intent modeling [13, 14]. Interactive intent modeling enhances human information exploration through computational modeling (visualized for interaction), helping users search and explore via user interfaces that are highly functional but not cluttered or distracting. In the area of user interface and semantic retrieval, a number of user interfaces have been designed to facilitate construction of queries for better searching results [15, 16]. These interfaces can help users to describe their intention more precisely and explicitly, but require users to use a terminology that is compatible with the data schema and background knowledge [17, 18]. An interactive interface is always provided to enhance the semantic searching [24–26].

Inductive Learning and Information Retrieval: Applying machine learning methods to information retrieval system has proved effective [19, 20]. Knowledge bases supporting stronger inferences about relationships between query and document concepts would enable learning in an IR system to be driven more from the knowledge base. Knowledge about the domain that documents and queries are concerned with is clearly useful to IR. As discussed in the Introduction section, solving relevance feedback in OWL based semantic retrieval can be naturally converted into an ILP problem without requiring any background knowledge for users. But there is not much work focusing on how to adapt ILP technology to describe user's intent and enhance searching result in OWL KB.

Probabilistic Information Retrieval Model and Uncertain Inference in Information Retrieval: In probabilistic IR models, we seek to estimate the probability that a specific document d_m will be judged relevant w.r.t. a specific query q_k, denoted as $P(R|q_k, d_m)$. There are two major distinguished approaches in probabilistic IR: The classical approach is based on the concept of relevance, that is, a user assigns relevance judgments to documents w.r.t. a query, and the task of IR is to yield an approximation of the set of relevant documents. The philosophy of this approach is used by many probability search methods [21, 22]. The new approach formulated overcomes the subjective definition of relevance judgment in an IR system by generalizing the proof-theoretic model of database systems toward uncertain inference. That's just the approach we adopted in our work. For applying this approach to OWL knowledge base, the core technique is to implement the IR-oriented uncertain inference, some work has focused on this topic, such as [23]. They provide us a practical way to implement the probability IR in OWL knowledge base.

3 Concept Learning and Probabilistic Information Retrieval

Concept Learning

We assume the reader to be familiar with the basic concepts from description logics and ILP and also fundamental OWL based Semantic Web technologies [27]. Concept learning in OWL knowledge base can be considered as concept learning in the description logics knowledge base, as defined in Definition 1.

Definition 1 (Concept Learning in Description Logics, CLDL). Let a concept name *Target*, a knowledge base K (not containing *Target*), and the sets E_+ and E_- with elements of the form *Target*(a) ($a \in I$, I is the set of individuals of K) be given. The learning problem is to find a concept C such that Target does not occur in C and for $K' = K \cup \{Target \equiv C\}$ we have

(1) $K' \models E_+$ (completeness)
(2) $K' \not\models E_-$ (consistency)

Such a concept C is called correct.

Adopting the CLDL to IR will lead to a four-step IR procedure: Step 1: Performing regular search on the knowledge base; Step 2: According to the hitting list, user inspects several top-ranked resources and marks those with relevance/positive(E_+) or irrelevance/negative(E_-)(optional); Step 3: Based on E_+, E_- (optional) and knowledge base K, system automatically learns the target concept of user's intention; Step 4: Performing the new search, that is finding the instances of target concept in the knowledge base. This procedure, called CLDL IR in the following, is repeated as long as there are new relevant documents near the top of the list or the user gives up.

Traditional Probabilistic IR Model

In most probabilistic IR models, the probability that a specified document d_m will be judged relevant w.r.t. a specific query q_k. In order to estimate this probability (denoted as $P(R|q_k, d_m)$), we regard the distribution of terms within the documents of the collection. The basic assumption is that terms are distributed differently within relevant and non-relevant documents. Let $T = \{t_1, \ldots, t_n\}$ denote the set of terms in the collection, then we can represent the set of terms d_m^T occurring in document d_m as a vector $x = (x_1, x_2, \ldots, x_n)$ with $x_i = 1$, if $t_i \ d_m^T$ and $x_i = 0$ otherwise.

A document represented by a binary vector x being relevant to a query q_k can be computed by the computing the odds of event(that a specified document d_m will be judged relevant w.r.t. a specific query q_k) [3]:

$$O(R|q_k, \vec{x}) = O(R|q_k) \prod_{x_i=1}^{n} \frac{P(x_i = 1|R, q_k)}{P(x_i = 1|\overline{R}, q_k)} \cdot \frac{P(x_i = 1|R, q_k)}{P(x_i = 1|\overline{R}, q_k)}$$

Now let $p_{ik} = P(x_i = 1|R, q_k)$ and $q_{ik} = P(x_i = 1|\overline{R}, q_k)$, for applying this model, we should estimate the parameters p_{ik} and q_{ik} for the terms $t_i \in q_k^T$. This can be done by means of relevance feedback. Let f denote the number of documents presented to the user, of which r have been judged relevant. For a term t_i, f_i is the number among the f documents in which t_i occurs, and r_i is the number of relevant documents containing ti. Then we can use the estimates $p_{ik} \approx r_i/r$ and $q_{ik} \approx (f_i - r_i)/(f - r)$. Some other better parameter estimation methods such as maximum likelihood estimate, Bayesian estimate are also popular ways to find the parameters.

Combining Concept Learning Based IR and Probabilistic IR Model

We extend CLDL IR to its corresponding probabilistic model. We open the standard in the step 4 of the CLDL IR procedure: Instead of finding the instances that strictly satisfy the concept learned, we find the instances that have relatively high chance to match the concept learned. The CLDL probabilistic IR can be considered as finding the instances set $D = \{d \mid$ there exists at least one annotation concept Ca of d that $P(R|Ct, Ca) > t\}$, where Ct is concept learned, t is a probabilistic threshold. So the core problem of combining concept learning and probabilistic IR in OWL knowledge base is to compute $P(R|Ct, Ca)$. Based on the semantic relations, $P(R|Ct, Ca) > t$ can be interpreted as $P(Ca \sqsubseteq Ct) > t$ or $P(Ct \sqsubseteq Ca) > t$. The most distinguished feature of the probabilistic IR model using concept learning is to interpret probabilistic IR as uncertain inference. Adopting uncertain inference can overcome some major short-comings of traditional probability IR models from relevance computing perspective. Table 1 is a brief comparison of the two different probabilistic IR models.

Table 1. A brief comparison of probabilistic IR model on vector space model and probabilistic IR model on concept learning in OWL KB

	Probabilistic IR model on vector space model	Probabilistic IR model on concept learning in OWL KB
Document d_m representation	$d_m = (x_1, x_2, \ldots, x_n)$	$d_m{:}C_1, \ldots, d_m{:}C_k$ ($d_m{:}C$ means d_m is annoted with $C_{k)}$
Query presentation	$q_k = (y_1, y_2, \ldots, y_n)$	C_q (Concept learned from feedback)
Query processing	Similarity Computation	Instance checking
Relevance feedback processing	Estimate parameters	Learning new concept from feedback
Relevance computation	Bayes' theorem or other	Computing $P(R|Ct, Ca)$ based on uncertain inference

4   Probabilistic IR in OWL KB Using Concept Learning

Constructing Probabilistic Knowledge Base

For constructing a probabilistic knowledge base, we apply the distribution semantics of probabilistic logic programming to description logics [28]. A program following this semantics defines a probability distribution over normal logic programs called *worlds*. Then the distribution is extended to concept learned and the probability of a query is obtained by marginalizing the joint distribution of the concept and the programs.

A probabilistic knowledge base K contains a set of probabilistic axioms which take the form $p :: E$, where p is a real number in $[0, 1]$ and E is a description logic axiom. The idea is to associate independent Boolean random variables to the probabilistic axioms. To obtain a *world* w we decide whether to include each probabilistic axiom or not in w. A *world* therefore is a non-probabilistic KB that can be assigned semantics in the usual way. A query is entailed by a *world* if it is true in every model of the *world*.

The probability p can be interpreted as an epistemic probability, i.e., as the degree of our belief in axiom E. For example, a probabilistic concept membership axiom $p ::$ $a : C$ means that we have degree of belief p in $C(a)$. A probabilistic concept inclusion axiom of the form $p :: C \sqsubseteq D$ represents the fact that we believe in the truth of $C \sqsubseteq D$ with probability p.

Assigning Probability to Axioms in KB

It will be handy if the concepts and axioms in knowledge base have been assigned a probability by domain experts or in other way. Since the knowledge base we are working on and most knowledge bases have not been augmented with probabilities, we need find a way to construct a probabilistic knowledge base to support probabilistic IR. The axioms we concern most in IR usually have the form of subsumption between concepts, such as $C \sqsubseteq D$ or $\forall r.C \sqsubseteq \forall r'.C$, so we divide the probability assigning task into two parts: assigning probability to subsumption relations explicitly asserted in the knowledge base, and assigning probability to subsumption relations between non-atomic concepts.

The subsumptions between atomic concepts are explicitly defined by domain experts in knowledge base, which usually have the form $C \sqsubseteq D$, where C and D are atomic concept. So it is obvious and comprehensible to assign real value 1 to this kind of explicit subsumptions.

It becomes more complicated when assigning probability to the subsumptions between non-atomic concepts. The non-atomic concept could be any legal concept constructed with constructors and atomic concepts in knowledge base, so there does not exist explicit subsumption assertions between them. Now we need to compute the probability of this kind of subsumption based on the known relations.

According to the semantics of description logics, $C \sqsubseteq D$ iff $C \sqcap \neg D$ is unsatisfiable in the knowledge base. For axioms with the form $C \sqsubseteq D$ not defined explicitly, the probability of $C \sqsubseteq D$, denoted as $p(C \sqsubseteq D)$, equal to $p(C \sqcap \neg D)$, which means we can assign axiom $C \sqsubseteq D$ with the probability of unsatisfiability of concept $C \sqcap \neg D$. In the context of model-theoretic semantics of description logic, the computation of $p(C)$, that represents the probability of any individual is an instance of concept C in the KB, can be solved by instance checking problem, that is

$$p(C) = |\text{Instance}(C)|/|\text{Instance}(\top)| \tag{1}$$

It is the ratio of number of instances of C to the number of all instances (represented by $\text{Instance}(\top)$) in the knowledge base. We use this formula to compute the probability of atomic concepts.

For computing the probability of compound concept such as $C \sqcap \neg D$, we employ the distribution semantics to solve the problem. The idea of this kind of computation is to acquire the probability of unsatisfiability of a compound concept w.r.t a knowledge base through standard description logics reasoning mechanism, explained as follows.

Let a couple (E_i, k) represent an atomic choice defined in distribution semantics, where E_i is the ith probabilistic axiom in KB and $k \in \{0,1\}$. k indicates whether E_i is chosen to be included in a world ($k = 1$) or not ($k = 0$). A composite choice ck is a consistent set of atomic choices, i.e., $(E_i, k) \in ck$, $(E_i, m) \in ck$ implies $k = m$. The probability of a composite choice ck is

$$p(ck) = \prod_{(E_i,1) \in ck} p_i \prod_{(E_i,0) \in ck} (1 - p_i) \tag{2}$$

where p_i is the probability associated with axiom E_i. A selection σ is a total composite choice, i.e., it contains an atomic choice (E_i, k) for every axiom of the theory. A selection σ identifies a theory w_σ called a world in this way: $w_\sigma = \{E_i | (E_i, 1) \in \sigma\}$. Let us indicate with S_K the set of all selections and with W_K the set of all worlds. The probability of a world w_σ is

$$p(w_\sigma) = p(\sigma) = \prod_{(E_i,1) \in \sigma} p_i \prod_{(E_i,0) \in \sigma} (1 - p_i) \tag{3}$$

$P(w_\sigma)$ is probability distribution over worlds, i.e., $\sum_{w \in W_K} p(w) = 1$.

Now we can assign probabilities to non-atomic concepts. Given a world w, the probability of a non-atomic concept Cp is defined as $p(Cp|w) = 1$ if $w \vDash Cp$ and 0 otherwise. The probability of a concept can be defined by marginalizing the joint probability of the concept and the worlds:

$$p(Cp) = \sum_{w \in W_K} p(Cp, w) = \sum_{w \in W_K} p(Cp|w)p(w) = \sum_{w \in W_K : w \vDash Cp} p(w) \qquad (4)$$

We noted that the computation of $p(C)$ in Eq. (1) and $p(Cp)$ in Eq. (4) depends on different semantic models of description logics. But it does not make sense for probability computation because it is feasible to map distributional to model-theoretic semantic spaces, so the two computing methods we adopted are compatible with each other and can work together for probability IR in OWL KB.

Algorithm for Concept Learning based Probabilistic IR in OWL KB

We can perform probabilistic IR as described in Algorithm 1.

Algorithm 1: concept learning based probability IR in OWL knowledge base

Input: a probabilistic OWL knowledge base K, user's feedback F from the previous search

Output: documents related to user's feedback

1. Ct = ConceptLearning(F, K);   // concept Ct learned from F and K represents the query intent

2. for each document d annotated with Ca in K // Using unsatisfiability to measure the relevancy of the d and query

3. p = Unsatisfiability($Ca \sqcap \neg Ct$);//computing the unsatisfiability of $Ca \sqcap \neg Ct$, that means d is more specific than query

4. p' = Unsatisfiability($Ct \sqcap \neg Ca$); //computing the unsatisfiability of $Ct \sqcap \neg Ca$, that means d is more general than query

5. Ranking the documents by $max(p, p')$;

In Algorithm 1, the concept learning method is CELOE (Class Expression Learner for Ontology Engineering) borrowed from Lemman's work [9]. Both concept learning method and unsatisfiability checking rely on the description logics reasoner to complete their procedures, so basically, the reasoner efficiency decides the complexity of the algorithm we proposed. In order to make the reasoning service practical for applications, we restrict the expressivity of description logic we used that has data complete for polynomial complexity in our work.

5 Experiment

We collected a total of 21,000 Web pages from WebMD.com, and selected 1,275 web documents covering 3 topics of ADHD, allergy, and anxiety-panic as evaluation pages, and design 2 query tasks for each topic as evaluation task. The correct answer of each query is identified manually and used for calculating the recall and precision. 600 concepts and 2350 relations (roles) derived from MeSH vocabularies and MeSH tree structure are defined in the complete *TBox*, we only consider the MeSH resources that are relevant to the pages we collect (at least 27,336 concepts will be defined if we use the whole MeSH descriptions). 21,000 instances of concepts and 86,000 instances of relations are asserted in the complete *ABox*. We annotated each of the 1,275 documents with atomic concepts defined in *TBox*, and assigned a probability p for each annotation manually. That is $p :: d : Ca$, where d is the document, Ca is the concept document d belongs to. Therefore a probability healthcare knowledge base was prepared well for implementing concept learning based probabilistic IR. All tests were run on a 2.4 GHz Core i7 core machine with 8 GB RAM.

5.1 Describing User's Intent

We will show how concept learned from feedback represent user's intent and how relevant a document related to concept learned. Tables 2, 3 and 4 illustrate the concept Ct representing user's intent and the relevance between a document annotated with concept Ca and Ct on three topics we chose respectively. Each table shows the results about a given topic. The columns of the table are the two concepts describing the search intent for two query tasks respectively, we only selected one concept from the concept learning results. Each row is the concepts we used for annotating a document, if there are more than one concept used for annotating, Ca is the conjunction of these concepts. The numbers is the quantification we used to judge the relevance of a document w.r.t. a query, which is the higher probability of the unsatisfiability of concept $C_a \sqcap \neg C_t$ and that of $C_t \sqcap \neg D_a$, as described in Algorithm 1.

Table 2. The relevance of a document w.r.t. a query in ADHD topic (Q1: AttentionDeficit $\sqcap$ Therapy; Q2: BrainDysfunction $\sqcap$ $\exists$ hasTopic. DeficitDisorder)

Concepts Ca used to annotate a document	Concept Ct used to describe a search intent	
	Q1	Q2
Minimal brain, Dysfunction	0.32	0.93
Disorders	0.29	0.94
ADDH	0.47	0.56
AttentionDeficit, Therapy	0.96	0.21
Hyperactivity, Attention	0.27	0.36
Brain, Pathology	0.08	0.19
ADHD, Genetics	0.16	0.28
Brain, Nursing	0.33	0.15
AttentionDeficit, Disorder	0.39	0.96
Brain, Disorders	0.12	0.96

Table 3. The relevance of a document w.r.t. a query in Allergy topic (Q1: Allergy ⊓ Immunology; Q2: Allergy ⊓ Specialty ⊓ ∃ hasTopic.Methods)

Concepts C_a used to annotate a document	Concept C_t used to describe a search intent	
	Q1	Q2
Immune system	0.14	0.05
Immunologic techniques	0.11	0.28
Allergy, immunology	0.97	0.39
Allergy	0.67	0.85
Immunology	0.48	0.05
Allergy specialty	0.72	0.98
Radiation immunology	0.05	0.03
AutoImmune disease	0.09	0.17
Immune system phenomena	0.18	0.05
Acquired ImmuneDeficiency	0.03	0.01

Table 4. The relevance of a document w.r.t. a query in Anxiety topic (Q1: Nervousness ⊓ ∃ hasTopic.Physiopathy ⊓ ∃ hasTopic.Therapy; Q2: NeuroticAnxiety ⊓ Disorders)

Concepts C_a used to annotate a document	Concept C_t used to describe a search intent	
	Q1	Q2
Anxiety Separation	0.33	0.79
Hypervigilance	0.27	0.36
AnxietyState, Neurotic	0.39	0.35
Disorders, Anxiety	0.45	0.92
Anti-Anxiety Agents	0.43	0.35
Nervousness	0.52	0.68
Anxieties, Nursing	0.35	0.58
Anxiety Diagnosis	0.78	0.96
Anxiety, Genetics	0.32	0.36
Anxiety, Pathology	0.95	0.35

Taking ADHD topic as example, we now give a brief explanation of the concepts describing user's intent. For query Q1, AttentionDeficit ⊓ Therapy is one of the most accurate concepts learned from feedback, which means the user was interested in the documents related to both attention deficit and Therapy about it. Concept BrainDysfunction ⊓ ∃ hasTopic.DeficitDisorder describes the intent about brain dysfunction the deficit disorder caused by it. Basically the concept name we learned is self-explainable; we can understand and describe user's intent explicitly and in a formal way. This explanation of concept learned also works for the other two topics shown in Tables 3 and 4.

5.2   Efficiency of Uncertain Inference in OWL KB for Probabilistic IR

Benefiting from the great improvement of the performance of description logics reasoners in recent years, we can try to apply the description logics inference to a real domain specific information retrieval system. Because we adopted probability to the knowledge base, that will slow the inference procedure during query processing. Taking the run time of computing probability of $a{:}C$ as example (the run time of computing probability of a is an instance of C), we illustrate the efficiency of uncertain inference in Fig. 1. The x-axis is the number of axioms involved in the computation, y-axis is the run time of the computing. For each topic, we constructed 7 queries with different probabilistic axioms requirements (for example, concept *Anxiety* is a simple query without probabilistic subsumption, non-atomic concept Nervousness ⊓∃ hasTopic.Physiopathology ⊓∃ hasTopic.Therapy is a complex query involving probabilistic subsumption). Compared with queries without probability axioms, queries with probability axioms cost much more time. It is obvious that the queries involving less probability axioms consume less running time. The results from 3 different topics are consistent in tendency about number of probability axioms and run time.

In real application environment, a query usually involves few probabilistic axioms because the length of concept related to user's intent are restricted in considering the concept comprehensibility. Although the run time in Fig. 1 are pretty high when numbers of probability is more than 3, this circumstance happened much less than usual.

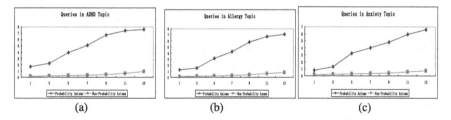

Fig. 1.   The run time of computing the probability of $a{:}C$ in different topics. (a) The run time in ADHD topic; (b) The run time in Allergy topic; (c) The run time in Anxiety topic.

Table 5 shows the effect of probabilistic TBox size on the query time. We generated enough axioms automatically for testing the efficiency(there exist many semantically redundant axioms in TBox). The probabilistic TBox size shows how many probabilistic axioms are defined in the TBox. For a query with form $a{:}C$, Table 5 shows query processing time of a standard TBox with 1000 axioms is 0.27 s, if 200 of this 1000 axioms are probabilistic assertions, the query processing time will be 0.95. The larger TBox and bigger probabilistic TBox size mean more query processing time. Similar to the experiment shown in Fig. 1, we did not compare the results with other related work. The reason for that is twofold: first, there is not a benchmark for IR-oriented uncertain inference by now, and constructing the dataset we used need much manual work. The second is the goal of this work is not to deal with the efficiency

Table 5. The query processing time(s) on different knowledge base TBox size

TBox axioms	Probabilistic TBox size				
	0	200	400	600	1000
1000	0.27	0.95	1.77	1.98	2.06
3000	0.74	2.67	4.02	4.46	4.68
5000	1.08	3.58	4.96	5.29	5.83

problem of uncertain inference in OWL KB but to test the applicability of the method proposed. So we only test the applicability of the method proposed on our dataset.

5.3   Overall Performance of the IR Prototype Designed

We developed a prototype of healthcare information retrieval system to prove the applicability of the CLDL probabilistic IR strategy. Figure 2 is the searching interface returned from the system for a query of "insect allergy": users can input the query keywords into the text box just as keyword based normal search, or input query by clicking the professional terms showed on the right of each result or the general descriptions on the left of the results. The distinguishing feature is the checkbox in front of the each searching result, users can mark the checkbox for expressing their positive intention of the current searching. The results displayed in this interface are only from one of the two collections with respect to the topic user concerned when formulating a query.

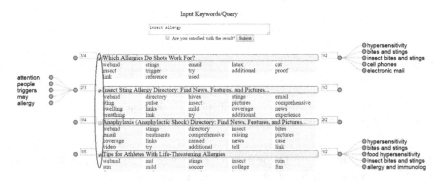

Fig. 2. An interactive searching interface for healthcare information retrieval

For each topic in the dataset, we divided the documents into two collections, one which is used for the initial query and relevance judgments, and the second that is then used for comparative evaluation. The performance of both initial query and feedback based query are compared to the second collection. The reason for that experiment plan is to pursue fairness in the evaluation. The concept learning based query processing based on the feedback from the initial query result. So it is obvious that we can start with an initial query and to compute a precision-recall graph. Following one round of feedback from the user, we perform the concept learning based query processing and again compute a precision-recall graph. It is straightforward to compare the two performances for evaluation, but it is unreasonable. Because the gains of the feedback based query are partly due to the fact that known relevant documents (judged by the user) are now ranked higher. Fairness demands that we should only evaluate with respect to documents not seen by the user.

We recruited 5 adults for the user study; each participant was required to find some information about the searching tasks devised for the experiment according to the content of the documents in the data collection. The participants completed the searching task through both normal searching interface and feedback supported searching interface. We recorded the query user formulated and relevant documents user marked on the interface for performing the CLDL probabilistic IR in the other collection of the topic user concerned. All the participants agree that the searching interface can support feedback in a simple but effective way and can help a lot to formulate a query that expresses their intentions correctly. Table 6 illustrates the average performance of normal semantic IR (S-IR in Table 6, which did not consider probabilistic reasoning during semantic IR procedure) and CLDL probabilistic IR(CL-PIR in Table 6) respectively according to an experimental record of 5 participants. For each topic, there are 2 tasks to be completed by querying the system. Since we only have small documents set for evaluation, the recall and the precision are expressed in fraction form instead of percentage form, the numerator and the denominator represent the number of documents corresponding to the classical formula of recall and precision in IR area respectively. Because the user's intent has been described explicitly and in a formal way, it is comprehensible that the improvement of recall and precision on CLDL probabilistic IR benefit from this.

Table 6. The performance of the normal semantic IR and CLDL probabilistic IR(recall and precision are in fraction form)

Topics/Query	Recall		Precision		Run time(s)	
	S-IR	CL-PIR	S-IR	CL-PIR	S-IR	CL-PIR
ADHD/Q1	16/16	16/16	16/23	16/18	0.42	1.84
ADHD/Q2	6/7	7/7	6/10	7/7	0.39	1.73
Allergy/Q1	25/28	27/28	25/36	27/29	0.47	2.26
Allergy/Q2	11/12	12/12	11/16	12/12	0.44	1.97
Anxiety/Q1	16/19	19/19	16/42	19/23	0.43	2.05
Anxiety/Q2	4/5	5/5	4/10	5/6	0.56	2.46

6 Conclusion

Supported by Semantic Web techniques, we combined concept learning techniques and probabilistic information retrieval model using uncertain inference in description logics knowledge base. Concept learning results are used to describe user's searching intent and the uncertain inference is used to judge the relevance of the document with respect to a query. The experiment results show the efficiency of the information retrieval strategy discussed in this work. The main advantage of the method proposed is that we can explain and describe user's search intent explicitly and in a formal way. The semantic information implied in the knowledge base is fully exploited when performing the uncertain inference for IR; this also helps to improve the recall and precision.

However, the disadvantages are also obvious. Applying concept learning method to information retrieval procedure needs much more preprocessing work, because we need a probability knowledge base. The inference efficiency in the knowledge base is the bottleneck of the ontology based application, so the method proposed cost more than non-inference based methods. It is difficult to quantify the experiment result in an objective way because there is not benchmark by now for testing probability information retrieval in description logics knowledge base. We only test our method on very limited resources of a local search service; it is still not easy to apply this method to a web-scale IR system. We will address this problem in our future research. The further work also includes adapting concept learning algorithm to fit the task of describing user's intent better and improving the efficiency of information retrieval-oriented uncertain inference.

Acknowledgement. Supported by the Fundamental Research Funds for the Central Universities (grant number GK201503066), and National Natural Science Foundation of China (grant number 61771297).

References

1. Bansal, M., Arora, J.: A review on ontology based information retrieval system. Int. J. Eng. Dev. Res. **4**(2), 263–265 (2016)
2. Tulasi, R.L., et al.: Ontology-Based Automatic Annotation: An Approach for Efficient Retrieval of Semantic Results of Web (2017)
3. Manning, C.D., Raghavan, P., Schtze, H.: Introduction to Information Retrieval. Cambridge University Press, New York (2008)
4. Ruotsalo, T., Jacucci, G., Myllymäki, P., et al.: Interactive intent modeling: information discovery beyond search. Commun. ACM **58**(1), 86–92 (2015)
5. McGuinness, D.L., Van Harmelen, F.: OWL Web Ontology Language Overview (W3C Candidate Recommendation 2003) (2015). http://www.w3.org/TR/owl-features/
6. Basu, A.: Semantic web, ontology, and linked data. Information Retrieval and Management: Concepts, Methodologies, Tools, and Applications: Concepts, Methodologies, Tools, and Applications, 24 (2018)
7. Gulwani, S., Hernandez-Orallo, J., Kitzelmann, E., Muggleton, S.H., Schmid, U., Zorn, B.: Inductive programming meets the real world. Commun. ACM **58**(11), 90–99 (2015)
8. Džeroski, S.: Relational data mining. Springer, Boston (2009)
9. Bühmann, L., Lehmann, J., Westphal, P.: DL-Learner—A framework for inductive learning on the Semantic Web. Web Semant. Sci. Serv. Agents World Wide Web **39**, 15–24 (2016)
10. Lehmann, J., Hitzler, P.: Concept learning in description logics using refinement operators. Mach. Learn. **78**(1–2), 203–250 (2010)
11. Munir, K., Anjum, M.S.: The use of ontologies for effective knowledge modelling and information retrieval. Appl. Comput. Inf. (2017)

12. Krishnamurthy, S., Akila, V.: Information retrieval models: trends and techniques. In: Web Semantics for Textual and Visual Information Retrieval, pp. 17–42. IGI Global (2017)
13. Marchionini, G.: Exploratory search: from finding to understanding. Commun. ACM **49**(4), 41–46 (2006)
14. Hu, J., Wang, G., Lochovsky, F., et al.: Understanding user's query intent with wikipedia. In: Proceedings of the 18th International Conference on World Wide Web, pp. 471–480. ACM, (2009)
15. Zenz, G., Zhou, X., Minack, E., et al.: From keywords to semantic queries—incremental query construction on the Semantic Web. Web Semant. Sci. Serv. Agents World Wide Web **7**(3), 166–176 (2009)
16. Caruccio, L., Deufemia, V., Polese, G.: Understanding user intent on the web through interaction mining. J. Vis. Lang. Comput. **31**, 230–236 (2015)
17. Bobed, C., Esteban, G., Mena, E.: Enabling keyword search on Linked Data repositories: an ontology-based approach. Int. J. Knowl. Based Intell. Eng. Syst. **17**(1), 67–77 (2013)
18. Loggie, W.T.H.: Using inductive logic programming to assist in the retrieval of relevant information from an electronic library system. In: Notes of the Workshop on Data Mining, Decision Support, Meta Learning and ILP held at The Fourth European Conference on Principles of Data Mining and Knowledge Discovery, Lyon, France (2000)
19. Li, H., Zhengdong, L.: Deep learning for information retrieval. In: Proceedings of the 39th International ACM SIGIR conference on Research and Development in Information Retrieval. ACM (2016)
20. Lewis, D.D.: Learning in intelligent information retrieval. In: Machine Learning: Proceedings of the Eighth International Workshop (2014)
21. Fuhr, N.: Probabilistic models in information retrieval. Comput. J. **35**(3), 243–255 (1992)
22. Sontag, D., et al.: Probabilistic models for personalizing web search. In: Proceedings of the Fifth ACM International Conference on Web Search and Data Mining. ACM (2016)
23. Zese, R., Bellodi, E., Lamma, E., Riguzzi, F., Aguiari, F.: Semantics and inference for probabilistic description logics. In: Bobillo, F., et al. (eds.) Uncertainty Reasoning for the Semantic Web III, URSW 2012, URSW 2011, URSW 2013. LNCS, vol. 8816, pp. 79–99. Springer, Cham (2014). https://doi.org/10.1007/978-3-319-13413-0_5
24. Ferré, S., Hermann, A.: Semantic search: reconciling expressive querying and exploratory search. In: The Semantic Web–ISWC 2011, pp. 177–192 (2011)
25. Varožek, M.: Exploratory search in the adaptive social semantic web. Inf. Sci. Technol. Bull. ACM Slovakia **3**(1), 42–51 (2011)
26. Zenz, G., Zhou, X., Minack, E., et al.: Interactive query construction for keyword search on the semantic web. In: De Virgilio, R., Guerra, F., Velegrakis, Y. (eds.) Semantic Search over the Web, pp. 109–130. Springer, Berlin Heidelberg (2012)
27. Baader, F.: The Description Logic Handbook: Theory, Implementation and Applications. Cambridge University Press, New York (2003)
28. Herbelot, A., Vecchi, E.M.: Building a shared world: mapping distributional to model-theoretic semantic spaces. In: EMNLP (2015)

Diabetic Retinopathy Classification Using C4.5

Mira Park[(⊠)] and Peter Summons

School of Electrical Engineering and Computing, University of Newcastle,
Callaghan, NSW 2308, Australia
{mira.park,peter.summons}@newcastle.eud.au

Abstract. Early detection of diabetic retinopathy (DR) can prevent blindness and improve the quality of life. Practical detection requires a cost-effective screening over a large population. The presence of Microaneurysms (MAs) in a retinal image is the earliest sign of DR. This paper presents an efficient method to automatically detect MAs in a retinal image. The method is based on an advanced wavelet transform and the C4.5 algorithm (a categorization algorithm to distinguish DR and non-DR cases). It uses both the green and red channel data in RGB retinal images for detection of small sized MAs and obtains image feature parameters from the input image. A system was developed to implement the proposed method that displayed a sensitivity of 0.92 and a precision of 0.82.

Keywords: Diabetic retinopathy · Microaneurysms · C4.5
Automatic detection

1 Introduction

Diabetes is a systemic disease affecting all parts of the body. In particular, Diabetic Retinopathy (DR) is defined as the damage done to the eye due to diabetes and is a major source of blindness. Vision loss is often related to retinal blood vessel and neuronal damage. Compelling evidence suggests that inflammation may play an important role in diabetic related retinal vascular and neuronal degeneration.

The World Health Organization (WHO) predicts that the number of patients with diabetes will increase to 366 million in 2030 [1]. In patients with diabetes, early detection of DR is important, because treatment methods can slow down the progression of the disease. Early diagnosis and treatment have been shown to prevent visual loss and blindness.

The use of digital photography of the retina examined by expert readers during screening programs has been shown to be both sensitive and specific in the detection of the early signs of DR. Since the retina is vulnerable to microvascular changes of diabetes and DR is the most common complication of diabetes, eye fundus (back of the retina) imaging is considered a non-invasive and painless route to screen and monitor potentially diabetic eyes.

To increase the ability to screen large populations, several groups have proposed the use of automated computer systems for initial determination as to who (which screened patients) should be referred to an ophthalmologist for further testing and who can safely return for screening 1 year later [2]. Detecting DR is a time-consuming and

© Springer Nature Switzerland AG 2018
K. Yoshida and M. Lee (Eds.): PKAW 2018, LNAI 11016, pp. 90–101, 2018.
https://doi.org/10.1007/978-3-319-97289-3_7

manual process that requires a trained clinician to examine and evaluate digital color fundus photographs of the retina. For automated systems to be applied in clinical practice, they should be evaluated extensively and thoroughly. Evaluation should firstly show that automated systems can detect DR with a sensitivity comparable to that of a human expert, whilst still maintaining a high enough specificity to attain the needed reduction in the ophthalmologist's workload. Since diagnostic procedures require the attention of an ophthalmologist, as well as regular monitoring of the disease, the workload and shortage of personnel will eventually exceed the current screening capabilities. To cope with these challenges, digital imaging of the eye fundus, and automatic algorithms based on image processing and computer vision techniques offer great potential.

DR is a microvascular complication of diabetes, causing abnormalities in the retina and typically begins as small changes in the retinal capillaries. The smallest detectable abnormalities, microaneurysms (MAs), appear as small red dots in the retina and are local distensions of the weakened retina capillary (Fig. 1). The MAs are a tiny area of blood that extends out from an artery or vein in the retina at the back of the eyeball. The method proposed in this paper was used to develop an automatic retinal fundus image (color and laser-scanned) analysis system for the early detection of DR, through identification of MAs in the retinal image.

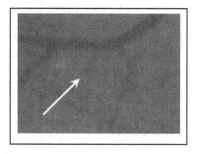

Fig. 1. Example of a microaneurysm (Color figure online)

Interest in automatic detection of DR has been increasing with the rapid development of digital imaging techniques and computing power. Some of the first automated detection methods for DR were published by Baudoin [3] to detect MAs from fluorescein angiograms. A considerable amount of effort has been spent on automated detection and diagnosis of DR during the past 10–15 years. For detection of DR various steps are normally carried out, these being: preprocessing, feature extraction and classification.

Some techniques have been proposed in the literature to enhance steps in the detection of MAs within digital images. For preprocessing, a morphological transformation [4] was proposed to make the dark lesions have low content in the green color plane.

The MAs typically appear as a small red dot in the retinal image so Mizutana et al. [6] employed a double-ring filter (with diameters of 5 and 13 pixels for the inner and outer rings respectively) to detect candidate regions for the MAs. Bernardes et al. [7] used a MA-tracker to count MAs. In other works a density function [4, 8] was defined to find the number of similar pixels in the neighbourhood of a selected pixel. When a target pixel was established, region growing could then be used to group similar pixels into the target lesions [5, 9, 10]. Morphological filtering could also be used to further enhance detection using 15 features, which includes the binary features, grey level and color features [11]. These algorithms produce candidate lesions and that are then classified into MAs or false positives using a rule-based method and an artificial neural network (ANN). Istvan Lazar [12] proposed an analysis of directional cross-section profiles centered on the local maximum pixels of the preprocessed image.

Unfortunately, these studies use their own data for evaluation and that data is not readily comparable to the high performance requirements necessary for application to public data. Several reasons can be considered for this: It is very hard to segment small regions, such as MAs, when the images are sourced from different locations using different equipment as the image formats and lighting conditions can vary. Even though the MA candidates are segmented, it is not easy to classify them into categories of true MAs and false MAs using the mean intensity of MA candidates. This is because the mean intensity of the MAs cannot be just one value for an input image because the color fundus image has a brighter intensity at the middle than the intensity at the side of an eyeball area. These issues become problematic when the training and testing images are from different sources (locations and equipment as indicated earlier). These are problems that need to be solved and so the method proposed in this paper attempts to address them in a system developed using the Isotropic Undecimated Wavelet Transform (IUDWT) and the C4.5 algorithm, a classifier to distinguish DR cases from non-DR cases.

2 Material

The proposed method and the resulting system described in this paper used datasets that (the authors expect) are typical of the populations on which the proposed system will be used. Although internally valid, this approach does not allow external validity (on other populations and datasets) to be determined. Many groups have evaluated components of DR screening systems on smaller datasets. Recently, more public data for the evaluation of algorithms has become available.

The DIARETDB1 database [14] was used to test the system. The 89 color fundus images are 1152 × 1500 pixels in size, and were captured using the same 50° field-of-view digital fundus camera with varying imaging settings. 84 of the images contain at least mild non-proliferative signs (MAs) of the DR and 5 are considered as normal. The images were annotated by 4 medical experts.

The database consists of two categories of image sets: a training set and testing sets. The training set has 29 images and the testing sets 60 images.

3 Methods

MAs are the smallest detectable abnormalities that the system will detect and are less than the diameter of the optic vein in size. They are isolated and appear as dark small red dots in the retina. In common with vessels, the best contrast for showing MAs against the retinal image background is in the green channel of the color fundus image. However, in the green channel, the presence of brighter lesions such as hard exudates is similar to the red lesions, thus the red channel is also used because there is less contrast between the bright lesions and the background.

3.1   Preprocessing

There are many preprocessing methods to enhance the red lesions, such as the Moat operator [15], however such preprocessing reduces the size of the red lesions. To correctly diagnose the early stage of retinopathy it is important to be able to detect very small red lesions, which can include lesions of less than 10 pixels in size. Thus, it is not worthwhile to use complicated preprocessing as it could remove very small red lesions.

The method proposed in the paper first applies an adaptive nonlinear diffusion algorithm [13] to remove noise from the image. The nonlinear anisotropic diffusive process has the good property of eliminating noise whilst still preserving the accuracy of edges. However, filtering depends on the threshold of the diffusion process and the threshold varies not only from image to image, but can also vary from region to region within an image. This problem compounds with intensity distortion and contrast variation. The proposed method tries to minimize the problems by applying the Central Limit Theorem to automatically determine and set the threshold for the adaptive nonlinear diffusion algorithm.

3.2   Segmentation of MA Candidates

MAs in retinal color fundus images are isotropic, thus the two-dimensional Isotropic Undecimated Wavelet Transform (IUWT) [16–18] is used to segment the candidate MAs. The implementation is simple as follows:

a. Set the spine coefficients for the filter

$$h_0 = [1, 4, 6, 4, 1]/16$$

b. Compute the transform and the sum of one or more requested wavelet levels: at each iteration i:
    i.   scaling coefficients c_i are computed by lowpass filtering to preserve the mean of the original signal;
    ii. wavelet coefficients w_i are computed by subtraction. Therefore there is a zero mean and encoded information corresponding to different spatial scales present within the signal.
c. Starting with a signal $c_0 = f$, where f is the original image, subsequent scaling coefficients are calculated by convolution with a filter h^i

$$c_{i+1} = c_i * h^i$$

where $h_0 = [1, 4, 6]/16$ is derived from the cubic B-spline and is the upsampled filter obtained by inserting $2^i - 1$ zeros between each pair of adjacent coefficients of h_0.

d. wavelet coefficients are then simply the difference between two adjacent sets of scaling coefficients

$$w_{i+1} = c_i - c_{i+1}$$

e. after the computation of n wavelet levels, reconstruction of the original image is obtained as the sum of the final set of scaling coefficients and all wavelet coefficients

$$f = c_n + \sum w_i$$

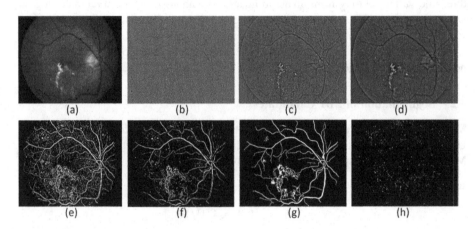

Fig. 2. Microaneurysm segmentation. (a) Input image (green channel) (b) reconstructed image by IUWT (wavelet level = 1) (c) reconstructed image by IUWT (wavelet level = 2) (d) reconstructed image by IUWT (wavelet level = 3) (e) segmented image (f) remove small objects (g) vessels segments (h) candidate MAs

Figure 2 shows three iterations of the IUWT. MAs are more clearly visible when the wavelet level is 3. From wavelet level 4 onwards, there is no improvement, thus we set the optimal wavelet level as 3. The reconstructed image (f) is segmented to produce candidate MAs through thresholding. It is useful to discriminate objects from the background in many image processing applications. An image containing an object, which has a homogeneous grey level and a background with a different grey level, usually possesses a bimodal histogram.

In a segmentation process, the value chosen for the threshold T is very important for medical applications. If too large threshold is chosen the number of false negative is increased, a very bad outcome as many true malignancies will be recognized as healthy. Medical applications may allow a grade number of false positives to ensure for minimal number of false negatives. We try to calculate an optimal threshold [18]. In the proposed method segmentation of candidate MAs is accomplished by the following:

a. Select a threshold, T

Otsu's method [18] produces optimal 1st and 2nd thresholding values in the range [0, 1] The 1st value thresholds the whole eye ball as the the eye ball is not masked from the background. Most MAs can be thresholded at (1st value + 2nd value)/1.5, so the initial T is set to be (1st value + 2nd value)/1.5

b. Segment the image using T

This produces two Groups: G1, pixels with value $\geq$ T and G2, pixels with value $<$ T

c. Take objects from *G1* (Fig. 2(e))
d. Remove all objects containing fewer than 5 pixels in *G1* (Fig. 1(f))
e. Fill the gaps between the close neighbours, or fill the holes in the objects

Only consider the gray level pixel value, as it can display any 'gaps', or 'holes', in the segmented objects

f. Divide the thresholded image into vessels, f_v and candidate MAs, f_{ma}

Compute an area (Ar) for each object. Assume there are n objects.
f_v ($Ar_i > 125$) = 1, $i = 1..n$ (Fig. 2(g))
f_{ma} ($Ar_i <= 125$) = 1, $i = 1..n$ (Fig. 2(h))
The vessel pixels have similar intensity to the MAs, thus the thresholded image contains both vessels and MAs. The MAs appear as small round dots whereas the vessel is represented by one large connected structure in the binary image. These two structures can be distinguished from each other based simply on their area.

Dark spots commonly occur on vessels where their contrast relative to the background reaches a peak. They can also occur at junctions, even where the vessels themselves are indistinct. Some candidate MAs may be segmented inside the nonhomogeneous vessels. The above process solves this problem.

3.3   Feature Space

There are three difficulties when trying to classify true MAs from candidate MAs.

Firstly, as Fig. 1 shows, the intensity of the MAs are darker than other structures, therefore the mean intensity of an object is a good property to use to classify the object as a MA. However, the mean intensity of the MAs cannot be just one value for an input image as the color fundus image has a brighter intensity at the middle than the intensity at the side of an eyeball area. The intensity of a MA can be defined by the intensity of the closest vessel. For the proposed system, the two closest vessels to the object are selected to set the object properties (see Fig. 3).

Secondly, the green channel is used to segment MAs because the MAs appear with their highest contrast in the green plane of the color image. The proposed system sometimes segments MAs from the green channel but it is not clearly visible on the original image's RGB channel and the segmented object can be just noise. This

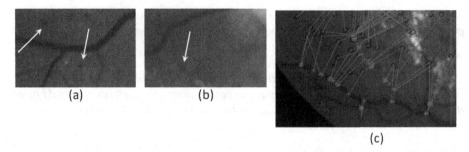

Fig. 3. MA with two closest vessel segments. (a) MA in the dark area (b) MA in the bright area (c) red circle: candidate MAs, green star: first close vessel, blue star: second close vessel (Color figure online)

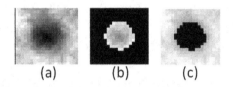

Fig. 4. Intensity difference between the object and background. (a) MA (b) Object only (c) background with offset = 5

problem is solved by applying a watershed algorithm for each object in the green channel. If the watershed algorithm still segments it the same as the IUWT, the object is a MA (see Fig. 4).

Thirdly, if the intensity of the object is too close to the background, it is not a MA. Therefore, the intensity around the object should be checked. The proposed system sets the offset as 5 pixels, and the intensity within the offset around the object is checked (see Fig. 5).

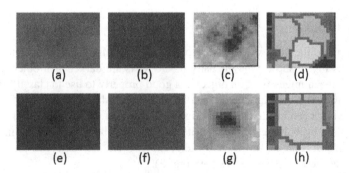

Fig. 5. Watershed segmentation for each object. *(a)–(d)*: false positive MA *(a)green channel (b) RGB (c) normalized image (d) result by watershed algorithm* (e)–(h): true positive MA *(e)green channel (f) RGB (g) normalized image (h) result by watershed algorithm* (Color figure online)

In order to build the feature space for classification of the objects, several measures are computed from each object. The system extracts the following features from each candidate MA.

In the thresholded image (black and white image):
Area: the number of pixels
Roundness: (perimeter of an object)^2/($4*pi*Area$)
Eccentricity: the eccentricity of the elliptical cross-section of the paraboloid
MajorA: the major axis

In the green channel:
meanIg: the mean intensity of an object
stdIg: the standard deviation of the intensity
d_mm_g: the difference between the maximum intensity and the minimum intensity of an object
vessell_1g: the mean intensity of the 1st closest vessel segment to the object
d_iv_1g: the difference between *meanIg* and *vessell_1g*
vessell_2g: the mean intensity of the 2nd closest vessel segment to the object
d_iv_2g: the difference between *meanIg* and *vessell_2g*
d_backgg: the difference between mean intensity of the background and an object

In the red channel:
meanIr: the mean intensity of an object
stdIr: the standard deviation of the intensity
d_mm_r: the difference between the maximum intensity and the minimum intensity of an object
vessell_1r: the mean intensity of the 1st closest vessel segment to the object
d_iv_1r: the difference between *meanIg* and *vesse5231l_1g*
vessell_2r: the mean intensity of the 2nd closest vessel segment to the object
d_iv_2r: the difference between *meanIg* and *vessell_2g*
d_backgr: the difference between mean intensity of the background and an object

3.4   Generation of Classification Rules

As the number of images is small and (=89), we are using C4.5 classifier to generate classification rules with relevant features. When using C4.5 decision tree, it is advantageous to limit the number of input features in the procedure of training tree in order to have a good predictive and small computationally intensive model. Because some of the features we acquire are little relevant with the final result of the tree. With a small feature set, the explanation of a rationale for the classification decision can be readily realized [23].

We first grow the tree, using a set of training data, quite often to its largest size. Secondly, we prune the tree to a smaller one by the excepted errors when testing this tree on the unseen cases. Finally, the classification rules would be generated from the tree with pruning. The three steps are as follows:

a. Constructing Decision Tree: If any algorithm can be said to as the foundation of this program, it is the process of generating an initial decision tree from a set of training cases.

b. Pruning Decision Tree: The initial one we obtained as above is a complicated tree that over fits the data by inferring more structure than is justified by the training cases. As many authors mentioned, most of all decision trees can benefit from simplification. In this case, we chose error-based pruning. When N training cases are covered by a leaf, E of them incorrectly, the resubstitution error rate of this leaf is E/N. For a given confidence level CF, the upper limit on the probability of expected errors on unseen cases can be found from the confidence limits from the binomial distribution, then

$$\text{Expected error} = U_{CF}(E, N) \times N$$

The main ideal: starts from the bottom of the tree and examines each nonleaf subtree. If replacement of this subtree with a leaf, or with its most frequently used branch, would lead to lower predicted errors, we prune the tree accordingly.

c. Classification rules are generated from the resulting tree after the pruning process.

4 Results

To evaluate the robustness of the system developed using the proposed method, a test set of 60 images from the DIARETDB1 [14] database. Between 0 and 59 MAs were present in each image, which had been manually annotated by 4 medical experts. Table 1 shows the results of processing the test set image data. The second column of Table 1 denotes the number of MAs manually annotated by the medical experts. The third column denotes the number of MAs detected by both the system and the experts (i.e., the True Positives: TP), the forth column denotes the number of MAs detected by the system but not by the experts (False Positives: FP) and the fifth column denotes the number of MAs missed by the system (False Negatives: FN). Table 1 also shows the sensitivity and the precision for each image in the sixth and seventh columns respectively, as well as the overall system sensitivity and precision (taken as the average over all images). The sensitivity of the system is taken as the number of true positives (real MAs detected) as a proportion of the total number of real MAs present in the images

Table 1. Results

Image No.	MAs	TP	FP	FN	Sensitivity	Precision
002	12	11	5	1	0.92	0.69
005	31	27	11	4	0.87	0.71
006	17	15	9	2	0.88	0.62
007	48	39	15	9	0.81	0.72
008	17	12	3	5	0.71	0.70
:						
089	5	5	7	0	1.00	0.71
Total	523	429	159	94	0.92	0.82

(= TP/(TP + FN). The precision of the system is given as = TP/(TP + FP). There are a total of 523 MAs in the test image set. The system detected 429 MAs and missed detecting 94 MAs. The system reported an extra 159 MAs that had not been confirmed by the experts. The system sensitivity is 0.92 and the system precision is 0.82.

Figure 6 shows the sensitivity and the precision for each image. An anomaly is noted in Image080, which has 3 MAs; the system detected 0 MA and 7 non-MAs, thus the sensitivity is 0 and the precision is 0. The dot with [0, 0] coordinator on the graph shows this case.

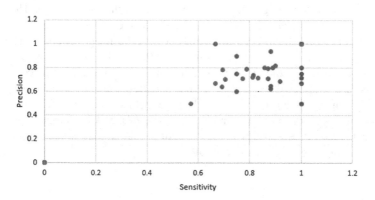

Fig. 6. Results: sensitivity and precision

Many other studies have used their own (non-public) datasets, or have simply categorised images as either containing microaneurisms or not, rather than locating the microaneurisms as the proposed method does. For a benchmark comparison of the proposed method with others that use the same public image dataset, Table 2 shows results from several other methods [20–22] using the same database for comparison of the sensitivity and precision of the proposed method.

Table 2. Sensitivity and precision method comparison

Method	Sensitivity	Precision
Purwita [21]	Not reported	Not reported
Ram [22]	88.46	Not reported
Manjaramkar [20]	80.06	97.5
Proposed Method	92	82

5 Conclusions

The prevention of DR concentrates on controlling the complications of diabetes in the eye through lifestyle and early treatment. These preventive actions can severely delay or stop the progression of the disease, prevent blindness and improve the quality of life [19].

The problem lies in the process of grading the eye fundus images which is time consuming and repetitive, and requires the attention of an ophthalmologist. Therefore, a cost-effective screening over large populations is required. Digital imaging of the eye fundus and automatic image analysis algorithms can provide a potential solution by reducing the workload of initial screening and quality assurance tasks.

An efficient method, based on an advanced wavelet transform and C4.5 algorithm, to automatically detect possible onset of DR by identifying the presence of MAs in a retinal image has been presented. It utilizes both the green channel and the red channel of RGB retinal images for diagnosis of early stage of retinopathy by detection of small size MAs and obtained image feature parameters from the input image. The method was used in a system and its robustness and accuracy in comparison to manual detection by medical experts was evaluated on a public image database. The results show the system is efficient, with a sensitivity of 0.92 and a precision of 0.82.

Among the many studies in the literature on computer analyses of color fundus images only a few utilize the detection of MAs and these use tailored data sets that may not allow the high efficiencies required when the methods are implemented on public data sets. This paper has presented an efficient system for MA detection using a public data set and shown that image processing of color fundus images has the potential to play a major role in early diagnosis of DR.

References

1. About diabetes: World Health Organization Report, 11 April 2015. www.who.int/diabetes/facts/en/index.html
2. Abramoff, M.D., Reinhardt, J.M., Russel, S.R., Folk, J.C., Mahajan, V.B., Niemeijer, M., Wuellec, G.: Automated early detection of diabetic retinopathy Ophthalmology, vol. 117, pp. 1147–1154 (2010)
3. Baudoin, C.E., Lay, B.J., Klein, J.C.: Automatic detection of microaneurysms in diabetic fluorescein angiography. Revue DEpidemiologie et de Sante Publique, pp. 254–261 (1984)
4. Grisan, E., Ruggeri, A.: Segmentation of candidate dark lesions in fundus images based on local thresholding and pixel density. In: The 29th Annual International Conference of the IEEE EMBS (2007)
5. Yang, G., Gagnon, L., Wang, S., Boucher, M.C.: Algorithm for detecting microaneurysms in low resolution color retinal images. In: Vision Interface, pp. 265–271 (2001)
6. Mizutani, A., Muramatsu, C., Hatanaka, Y.: Automated Microaneurysm detection method based on double-ring filter in retianal fundus image. IEEE Trans. Med. Imaging 7260, 1–5 (2009)
7. Bernardes, R., et al.: Computer-assisted microaneurysm turnover in the early stage of diabetic retinopathy. Ophthalmologica 223(5), 284–291 (2009)
8. Xu, X., et al.: Vessel boundary delineation on fundus images using graph-based approach. IEEE Trans. Med. Imaging 30(6), 1184–1191 (2011)
9. Spencer, T., et al.: Automated detection and quantification of microaneurysms in fluorescein angiograms. Graefe's Arch. Clin. Exp. Ophthalmol. 230, 36–41 (1991)
10. Zhang, B., Karray, K.: Microaneurysms detection via sparse representation classifier with MA and non-MA dictionary learning. In: IEEE International Conference on Pattern Recognition, pp. 1051–4651 (2010)

11. Breen, E.J., Jones, R.: Attribute openings, thinnings and granulometries. Comput. Vis. Image Underst. **64**(3), 377–389 (1996)
12. Lazar, I., Hajdu, A.: Retinal microaneurysm detection through local rotating cross-section profile analysis. IEEE Trans. Med. Imaging **32**(2), 400–407 (2013)
13. Jin, J.S., Wang, Y., Hiller, J.: An adaptive nonlinear diffusion algorithm for filtering medical images. IEEE Trans. Inf Technol. Biomed. **4**(4), 298–305 (2000)
14. Dibetic Retinopathy Database and Evaluation Protocol. www2.it.lut.fi/project/imageret/diaretdb1_v2_1
15. Sinthanayothin, C.: Image analysis for automatic diagnosis of diabetic retinopathy (1999)
16. Strack, J.L., Murtagh, F.: Astronomical image and signal processing. IEEE Signal Process. Mag. **18**, 30–40 (2001)
17. Olivo-Marin, J.C.: Extraction of spots in biological image using multiscale products. Pattern Recogn. **35**, 1989–1996 (2007)
18. Strack, J.L., Fadili, J., Murtagh, F.: The undercimated wavelet decomposition and its reconstructions. IEEE Trans. Signal Process. **1995**(16), 297–309 (2007)
19. Harper, C.A., Taylor, H.R.: Early detection of diabetic retinopathy. Med. J. Aust. **162**(10), 536–538 (1995)
20. Manjaramkar A., Kokare, M.: A rule based expert system for microaneurysm detection in digital fundus images. In: Computational Techniques in Information and Communication Technologies (ICCTICT) (2016). https://doi.org/10.1109/ICCTICT.2016.7514567
21. Purwita, A.A., Adityowibowo, K., Dameitry, A., Atman, M.W.S.: Automated microaneurysm detection using mathematical morphology instrumentation. In: Proceedings of the. 2nd International Conference on Communications Information Technology and Biomedical Engineering (ICICI-BME), pp. 117–120 (2011)
22. Ram, K., Joshi, G.D., Sivaswamy, J.: A successive clutter rejection based approach for early detection of diabetic retinopathy. IEEE Trans. Biomed. Eng. **58**(3), 664–673 (2011)
23. Akay, M.F.: Support vector machine combined with feature selection for breast cancer diagnosis. Expert Syst. Appl. **36**, 3240–3247 (2009)

Stock Price Movement Prediction from Financial News with Deep Learning and Knowledge Graph Embedding

Yang Liu[1], Qingguo Zeng[2], Huanrui Yang[3(✉)], and Adrian Carrio[4]

[1] Department of Industrial Engineering, Business Administration and Statistics,
Universidad Politécnica de Madrid, 28006 Madrid, Spain
yang.liu00@alumnos.upm.es
[2] South China Normal University, Shipai, Guangzhou, China
domceng@gmail.com
[3] Electrical and Computer Engineering Department,
Duke University, Durham 27708, USA
inociencio@gmail.com
[4] Center of Automatic and Robotic,
Universidad Politécnica de Madrid, 28006 Madrid, Spain
adrian.carrio@upm.es

Abstract. As the technology applied to economy develops, more and more investors are paying attention to stock prediction. Therefore, research on stock prediction is becoming a hot area. In this paper, we propose to incorporate a joint model using the TransE model for representation learning and a Convolutional Neural Network (CNN), which extracts features from financial news articles. This joint learning can improve the accuracy of text feature extraction while reducing the sparseness of news headlines. On the other hand, we present a joint feature extraction method which extracts feature vectors from both daily trading data and technical indicators. The approach is evaluated using Support Vector Machines (SVM) as a traditional machine learning method and Long Short-term Memory (LSTM) model as a deep learning method. The proposed model is used to predict Apple's stock price movement using the Standard & Poor's 500 index (S&P 500). The experiments show that the accuracy of news sentiment classification for feature selection achieved 97.66% by model of joint learning, the performance of joint learning is better than feature extraction by CNN, the accuracy of stock price movement prediction through deep learning achieved 55.44%, this result is higher than traditional machine learning. This model can give the investors greater decision support.

Keywords: Stock market · Deep learning · Event tuple · Financial news Knowledge graph embedding

1 Introduction

Stock market predictions are important business activities, today. However, establishing an accurate stock prediction model is still a challenging issue [1]. In addition to historical market movements, the prediction of stock market is also affected by

© Springer Nature Switzerland AG 2018
K. Yoshida and M. Lee (Eds.): PKAW 2018, LNAI 11016, pp. 102–113, 2018.
https://doi.org/10.1007/978-3-319-97289-3_8

financial news. Currently, financial news includes a large number of news events that are presented in the form of a knowledge graph [2]. As a result, both news events and historical market movements can improve the forecasting ability of the models.

Knowledge graphs were formally proposed by Google on May 2012 as an intention to improve the search engine's ability, enhancing the search quality and the users experience [3]. At present, due to the development of Artificial Intelligence technology for knowledge graphs, they have been widely applied in intelligent search, question and answering, and intelligent finance. A knowledge graph in finance aims to find the relationships in entities such as: companies, management, news events and user preferences [4]. These entities enable efficient, financial data-based decision making and provide business advice for investors to predict stock trends. In view of the above-mentioned reasons, this research focuses on investigating how to improve the accuracy of stock price predictions through the use of knowledge graphs.

Knowledge graphs are databases implementing semantic search by preserving the relationships between multiple entities [5]. The typical structure of a knowledge graph is: entity 1, relation, and entity 2. Entities are concrete things in the objective world. For example, "Tim Cook is the CEO of Apple" or "Apple is an Internet company", etc. Events are objective activities [6]. For example: the increase or fall of stocks, the release of new products, etc. The relation describes the objective relationship between concepts, entities and events. For example: "Jobs and his partners founded Apple in 1976", and "Apple released iPhone X in September 2017", etc. According to the definition of tuples in a knowledge graph, we can introduce the definition of an event tuple as a tuple (A, P, O), where A represents agent, P represents predicate, O represents object [7]. For example, "Apple says initial quantities of iPhone 7 Plus sold out". Event tuples group together relevant elements and can be used as an efficient way to improve prediction accuracy.

In the previous works [6, 7], there is a common problem in the integration of event tuple with knowledge graph. These works didn't not capture the structural information in the text, and these information is very important for affecting stock to increase or decrease. The bag of word and event tuple are used in previous work [6, 8], for example, Samsung sues Apple for stealing patents. If this text is represented in the bag of word, it could be represented that "Samsung", "sue", "Apple", "stealing", and "patents". Because there is no structured information in this text, this way is difficult to determine which company's stock price will increase and which company's share price will decrease. If this text represented by a structured event:(Agent: "Samsung"), (Predicate, "sue"), (Object: "Apple"). Although the object of each tuple is clearly known, it lost more information when text converted into a structural vector. Therefore, we propose a joint learning model of tuple and texts to maximize the retention of structured information in event tuple. Moreover, we select Apple's financial news as structure information extraction. We found that the accuracy of the classification reached 97.6% on this data. The structured information in the text was retained, these information more directly predicts the stock market.

2   Related Work

2.1   Deep Learning in Stock Market Prediction

There are three major applications of deep learning in stock market prediction. Firstly, the prediction of the operation in financial markets. Forecasting stock volatility and price of financial assets has attracted a lot of interest in the field of finance, especially in the prediction of price trend and direction [9]. As an example, in the work of [10], deep neural networks analyze the close price to predict the daily fluctuation of the S&P 500 index. Secondly, the application of natural language processing for forecasting Stock Market. In the work of [6], fully-connected neural networks and convolutional neural networks are used to build models that extract event tuples from the news. This method analyzes the impact of long-term events on the stock prices and judges on the future direction of the stock market price. Finally, deep learning can help investors improve their trading strategies. In the work of [11], deep learning is used to encode stock market information to design an effective portfolio based on the results of the decoding obtained. This method can screen market common factors and stock individual factors. None of the above methods make use of event tuples.

2.2   Knowledge Graph Embedding

Knowledge graph embedding is a type of representation learning between entities and relations in a knowledge base. The entities and relations are mapped into a low-dimensional space representing the semantic information between entities and relationships. Currently, there are some algorithms which explore knowledge graph embedding in translation distance models. Both entities and relations can be represented as vectors in the same space. Given a fact (h, r, t), the relation is interpreted as a translation vector r so that the embedded entities h and t can be connected by r with low error, for example, when $h + r \approx t$, the result (h, r, t) is true. Currently, there are different translation distance models. TransE [12] model is a computationally efficient and predictive model, this model that can be modeled well for "one-to-one" relational types. TransH [5] model, the head and tail vectors are mapped onto the hyperplane where the relationship is located, then the translation process is completed on the hyperplane. TransR [13] model is based on TrasnE model, which model's entities and relations are in different dimensions of space, this model defines one matrix for each relation, which is used to transform the entity vectors into the spaces, and this model completes the translation in the relational vector space. The TransD [14] model is an improvement based on TransR model, TransD model considers the transformation matrix based on TransR model, which is dynamically determined by the entity relationship. Comparing in these models, we need a simple mapping of one-to-one entities for extracting the feature, the accuracy of representation between entities and relationships is improved, the TansE model have these features. Therefore, the joint learning model of text from news and events tuple is presented based on TransE model in this work.

2.3   Representation Learning Based on Text and Knowledge

Because of the TransE model includes a one-to-one relation between two entities, in order to build a large-scale knowledge graph embedding, the relationship between many entities has to be constantly added. Recently, several methods have been applied in knowledge graph completion. In the work of [5] a word embedding was extracted from Wikipedia and a knowledge base was trained using the TransE model. This model makes the word representation corresponding to the entity in the text as close as possible to the entity representation in the knowledge base. Moreover, a deep architecture was proposed using both structural and textual information of the entities [15]. Specifically, three neural models were used to encode the valuable information from the entity in the text description. The method proposed by [16] illustrates two encoders for encoding entity descriptions: a continuous bag-of-words and a deep learning model. Previous works didn't make full use of description information in the text and a lot of semantic information was lost during the feature extraction process. In order to retain and fully extract the semantic information in financial news, we propose to extract feature vectors from the news text by means of a CNN model.

3   Task Description

3.1   Research Architecture

Figure 1 shows the research architecture [17]. The steps are as following:

1. Corpus collection and stock data compilation: By looking up keywords such as "Apple" at Thomson Reuters' website and writing a web-crawler to obtain Apple's financial news.
2. Data pre-processing: including corpus analysis, text normalization, word tokenization, label tagging and word to vector transformation.
3. Feature selection: selecting the features from the embedding and stock data layers to compute eigenvalues, to afterwards generate a feature vector using deep learning.
4. Deep Learning: Feature extraction, prediction model built with Tensorflow [18].
5. Results evaluation: analysis of the results and extraction of conclusions.

3.2   Dataset Description

In this work, we build our own financial news corpus with headlines from Apple, published in Thomson Reuters between October 2011 and July 2017. This database consists of 6,423 financial news headlines, each including its title and release date. The title is used for event embedding and feature extraction, while the release date is used to align the corresponding the financial news with trading data from a temporal series. As previous work has shown, using the headline can help reducing noise in text mining [19]. It has been shown that the headline concisely represents the content of the text [20]. We use exclusively the news headlines from Apple for stock price movement prediction.

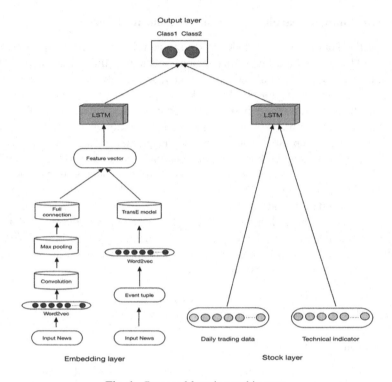

Fig. 1. Proposed learning architecture.

We also collected related daily stock data from the Standard & Poor's 500 (S&P 500) index in Yahoo Finance, in the same period as the stock data and the financial news headline from AAPL stock data. Daily trading data and technical indicator features are used by our model [21, 22]. And in the previous work [23], the technical indicators strongly prove the improvement of the stock forecast. Meanwhile, three technical indicators were used, which are calculated based on the AAPL daily trading data. the stock data of AAPL on S&P 500 from October 30, 2011 to July 8, 2017. In total, there are n = 1467 trading days. The variables are as follows:

1. Opening price; 2. Closing price; 3. High price; 4. Low price; 5. Volume; 6. Stochastic oscillator (%K); 7. Larry William (LW) %R indicator; 8. Relative Strength Index (RSI). The data is therefore arranged in 8-dimensional vectors.

The target output consists of a binary variable. A value of 1 indicates that the close price in the $t + 1$ will be higher than day t, while a value of 0 indicates that the close price drops down as compared to that of day t. Because of each news headline is presented as an event tuple, in Table 2 these headlines need to be transformed to event triplets, which has been done here using Reverb[1], there are 2,799 financial news headlines after filtering. The news event tuples and stock data are aligned creating input-output pairs, and those days without released news are left out. In Table 1,

---

[1] http://reverb.cs.washington.edu/.

the matches found are 941 pairs of event tuples and stock data. For this dataset, we use 80% of the samples for training and the remaining 20% for testing.

Table 1. Dataset information

Dataset	Total	Training	Test
Time interval	1.10.2011–30.7.2017	1.10.2011–12.30.2016	1.1.2017—7.31.2017
Event tuples	941	780	161

3.3   Data Pre-Processing

A web crawler to capture financial news has been developed. The obtained data consists of unstructured information, so the following procedures were applied to structure it:

1. Remove redundant information in the text, such as: stop words, excessive punctuation and repeated words.
2. Tagging the label for each news headline. The label consists of a categorical value among five possible labels: 0 (extremely negative), 1 (negative), 2 (natural), 3 (positive) and 4 (extremely positive). The label is given by the level of sentiment [24].

In Table 2, label 0 means that the event happened to an Apple's competitor, label 1 means that Apple lost something in this event, label 2 means that this event had nothing to do with Apple or that it did not cause any impact to Apple, label 3 means that this event caused the Apple to obtain something and label 4 means that Apple increased its profit or created more value from this event.

Table 2. Samples extracted from the dataset

Date	News headline	News headline event tuple	Label	Number of labels
25/1/2012	Motorola sues Apple for patent infringement	Motorola, sues, Apple for patent infringement	0	502
3/2/2012	Apple stops selling some devices online in Germany	Apple, stops selling, some devices	1	537
15/2/2012	Exclusive: Proview says any ban of iPad exports hard to impose	Proview, say, any ban of iPad exports	2	661
29/2/2012	How Apple, and everyone, can solve the sweatshop problem	Apple, solve, the sweatshop problem	3	692
2/4/2012	Apple's iPad tops Consumer Reports' list despite heat issue	Apple's iPad, tops, Consumer Reports' list	4	405

3. Word vector transforming. We use the Word2vec[2] model to train the word embedding. Word2vec is an algorithm for learning this distributed word vector, this vector is a real-size vector of fixed size, we determined its size, such as 300 dimensions. The word embedding was trained on 100 billion words from Google News, using the continuous bag-of-words architecture.

4 Methodology

4.1   Feature Selection by the Model of Joint Learning

4.1.1   Feature Extraction from the News Title Using CNN Model

As extracting the tuple from the original text may cause a loss of information, we also encode the original text of each piece of news' headline directly using a CNN model. Given the sequence of words in the title of a financial piece of news, we embed each word using Word2vec [25] as $(x_0, x_1, x_2 \ldots \ldots, x_n)$, and then concatenate these vectors as a matrix for the input to the CNN model. In our implementation, the joint learning model is made up of four consecutive layers: the first layer is the input layer, the second layer is a convolutional layer, the third layer is a max-pooling layer and the fourth layer is a fully connected layer for extracting features describing the relations between words. The convolution layer and max-pooling layer are designed according to TextCNN [26], which has been proved effective for emotion classification. Another linear layer followed by a softmax layer is attached to the fourth layer, which classifies the title into 5 previously defined emotional classes. The CNN model is trained using these emotion labels, with the classification loss denoted as E_d.

4.1.2   Feature Extraction from the Event Tuple Using TransE Model

In order to use the knowledge graph information and event tuple in news text at the same time. Given the knowledge graph $KG = (E, R, T)$, for each event tuple (h, r, t) $\in T$, h_s and t_s denote the structure vector representations of the head entity and tail entity the joint learning model is defined as the average of each word's word vector in the entity, obtained with the Word2Vec model. The two entity vectors are then mapped into the same relation space using a trained low rank weight matrix, i.e.

$$H = L_r h_s$$

$$T = R_r t_s$$

As assumed in TransE model, the relationship vector R should satisfy $H + R \approx T$, such as *Tim Cook + found ≈ Apple and Apple + release ≈ Iphone*8. The loss function of this structure model is defined as follow:

$$E_S = \|H + R - T\|_2^2$$

---

[2] https://code.google.com/archive/p/word2vec/.

4.1.3   Combined Loss Function for Feature Extraction

We combine two types of representation learning together to map the news titles into feature vectors. Here we denote the parameter set as $\emptyset = (L_r, R_r, \theta)$, where L_r, R_r are the mapping matrices for entities in the structure model, and θ are the weights of the CNN. For the structure model, we use a relationship vector R, identical to the result of the feature extraction layer of the CNN, to compute the loss:

$$E_S = \|H + R - T\|_2^2$$

And then combine this loss with the classification loss of the CNN model using L2 regularization to obtain the following overall loss function for feature extraction:

$$E = E_s + \alpha E_d + \beta \|\phi\|^2$$

4.2   Stock Market Prediction Model

4.2.1   Long Short-Term Memory Networks

On top of the feature extraction models, there are two LSTM models used in parallel, one is to interpret the output of the feature vector, and other one is interpreted the stock data features [17]. Long Short-Term Network is a special type of RNN that learns long-term dependencies. LSTM has already been proved successful in predicting stock prices [27].

LSTM model is an important part of deep learning [28]. Figure 2 shows a workflow describing the principles of this model. There are three gates: forget door, input door and output door. The door consists of a sigmoid activation function and a point-by-point multiplication operation. The model holds the hidden state of the previous time step, which has a three-wise connection: first one to the forget gate, f_t, second one to the input gate, i_t, and third one to the output gate, O_t.

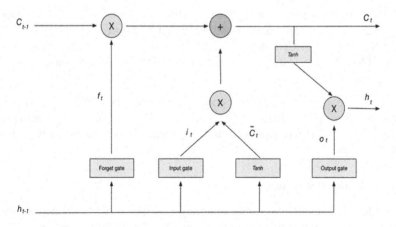

Fig. 2. Proposed LSTM model architecture.

In this case, i_t decides when to pass the activation into the memory cell, and O_t is used to activate the outgoing memory cell. Accordingly, for post-delivery, the O_t decides when to let the error flow into the storage cell, i_t decides when to let it flow out of the storage cell. We define the cell's new state as follow:

$$C_t = f_t * C_{t-1} + i_t * \tilde{C}_t$$

The advantage is that this method solves the vanishing gradient problem, which is caused by the gradual reduction of gradients during backpropagation. During training, we adjust several parameters in the LSTM. We consider the word embedding, originally sparse and defined over high-dimensional vectors. The word embedding eigenvectors are reduced to 50-dimensional vectors. These feature vectors are entered together with the stock data into the LSTM model.

4.2.2 Output Layer

The output of this model is a traditional fully connected layer with Logistic regression-based classification, that finds the probability distribution of eigenvectors over the labels.

5 Experiment and Results

5.1 Experiment Settings

In the experiments, to evaluate the influence of using both embedding layer and stock layer on stock price movement, a comparison is made between the hybrid model proposed in this work. We compare different models and the following notation identifies each model:

(1) **T-SVM:** Input of Tf-idf algorithm feature extraction and SVM prediction model [29].
(2) **J-SVM:** Input of joint learning feature extraction and SVM prediction model.
(3) **C-SVM:** Input of CNN feature extraction and SVM prediction model.
(4) **C-LSTM:** Input of CNN feature extraction and LSTM prediction model [17].
(5) **J-LSTM:** Input of feature extraction by the model of joint learning and LSTM prediction model.

The time step of above model set is one day, which makes a prediction of the next day. We compare the predictive performance on the test dataset in predicting the movement for the next day and evaluate the performance of the model by means of accuracy, F1-score [30].

5.2 Results and Discussion

We select feature vectors in the embedding layer by means of a joint learning model. When news is tagged with five sentiment labels, the classification accuracy is 97.66% for the five labels. This result demonstrates that the mapping ability of the head entity and tail entity in the same space is improved and the sparseness of each news and event tuple are reduced. In consequence, the joint learning model has a good performance using the feature vectors built from event tuples and financial news, and we consider the short-term interval of one day.

According to the results shown in Table 3, the accuracy of T-SVM, J-SVM and C-SVM is 49.17%, 54.92% and 45.6%, respectively. However, through joint learning using the proposed feature extraction model, the accuracy of C-LSTM reaches 51.32% and the accuracy of J-LSTM reaches 55.44%. Obviously, the accuracy of the joint learning model is higher than the other feature extraction models. In the feature selection by joint learning and CNN, the accuracy of C-SVM is 5.76% lower than the accuracy of C-LSTM, this decrease proves that feature extraction performance by means of joint feature learning is improved with respect to the Tf-idf algorithm and CNN. Comparing between traditional machine learning and deep learning techniques, the prediction accuracy using deep learning is 6.27% higher than traditional machine learning. Furthermore, F1-score increases from 53.48% to 71.33%. Thus, the comparison result provides the evidence that deep learning techniques can outperform traditional machine learning in stock price movement prediction, and the performance of feature selection by the joint learning surpasses the feature selection by CNN in this work.

Table 3. Accuracy and F1-score in the prediction of next day's close price movement

Predictive analytics Method	Accuracy	F1-Score
1. T-SVM	49.17%	44.58%
2. J-SVM	54.92%	53.48%
3. C-SVM	45.60%	33.96%
4. C-LSTM	51.32%	63.04%
5. J-LSTM	**55.44%**	**71.33%**

In the previous work [6, 17], the financial news headline as input that is difficult to investigate the relationship in some companies, the financial news from many competing companies may be noise data, these data reduce the stock's forecasting accuracy. The knowledge graph can provide the attributes and relationships of the entities and obtain more information from the companies, which increases the prediction accuracy of the stock prices [7]. The above accuracy results comply with the reported rates in previous works [20, 27]. Deep learning remains the preferred choice for stock forecasting models. LSTM model requires a lot of data for training. Given the scarcity of data in our problem, Deep learning would have difficulties to capture temporal features and the features of word embedding for long-term predictions. However, these experiments show that the model deep learning and joint learning is valid and provides good performance in predicting the movement of next day's price.

6 Conclusions and Future Works

In this paper, we propose the application of joint learning of event tuples and text for stock prediction, which solves the problem of text sparsity in feature extraction. It has been proved that our deep learning model predicts better than traditional machine learning in the short term. This can give investors feedback to support business decisions and to improve investment planning.

In future research, we will expand the developments in this paper. Firstly, our research will consider the incorporation and comparison of more machine learning models. Secondly, we will include companies from different areas, such as Boeing or Walmart and multiple financial news sources. The objective will be to prove the feasibility and generalization capability of the model. Finally, with respect to individual event tuple, the specific impact on the stock market needs to be analyzed and classification could be done over other time periods.

Acknowledgments. The authors Yang Liu would like to thank all the reviewers for their insightful and valuable suggestions. This work is supported by the China Scholarship Council (CSC).

References

1. Nguyen, T.H., Shirai, K., Velcin, J.: Sentiment analysis on social media for stock movement prediction. Expert Syst. Appl. **42**, 9603–9611 (2015)
2. Liu, K., Zhang, Y.-Z., Ji, G.-L., Lai, S.-W., Zhao, J.: Representation learning for question answering over knowledge base: an overview n ature. Acta Autom. Sin. **42**(6), 807–818 (2016)
3. Qiao, L., Yang, L., Hong, D., Yao, L., Zhiguang, Q.: Knowledge graph construction techniques. J. Comput. Res. Dev. **53**(3), 649–652 (2014)
4. Wang, Q., Mao, Z., Wang, B., Guo, L.: Knowledge graph embedding: a survey of approaches and applications. IEEE Trans. Knowl. Data Eng. **29**, 2724–2743 (2017)
5. Wang, Z., Zhang, J., Feng, J., Chen, Z.: Knowledge graph embedding by translating on hyperplanes. AAAI Conf. Artif. Intell. **14**, 1112–1119 (2014)
6. Ding, X., Zhang, Y., Liu, T., Duan, J.: Deep learning for event-driven stock prediction. In: IJCAI International Joint Conference on Artificial Intelligence, pp. 2327–2333 (2015)
7. Ding, X., Zhang, Y., Liu, T., Duan, J.: Knowledge-driven event embedding for stock prediction. Coling **2016**, 2133–2142 (2016)
8. Filliat, D.: A visual bag of words method for interactive qualitative localization and mapping. In: Proceedings of the IEEE International Conference on Robotics and Automation, pp. 3921–3926 (2007)
9. Harris, G.: A Survey of Deep Learning Techniques Applied to Trading (2016)
10. Xiong, R., Nichols, E.P., Shen, Y.: Deep Learning Stock Volatility with Google Domestic Trends. 2, 0–5 (2015)
11. Heaton, J.B., Polson, N.G., Witte, J.H.: Deep Learning in Finance, pp. 1–20 (2016)
12. Bordes, A., Usunier, N., Weston, J., Yakhnenko, O.: Translating embeddings for modeling multi-relational data. Adv. NIPS. **26**, 2787–2795 (2013)
13. Lin, H., Liu, Y., Wang, W., Yue, Y., Lin, Z.: Learning entity and relation embeddings for knowledge resolution. Procedia Comput. Sci. **108**, 345–354 (2017)

14. Ji, G., He, S., Xu, L., Liu, K., Zhao, J.: Knowledge graph embedding via dynamic mapping matrix. In: Proceedings of the 53rd Annual Meeting of the Association for Computational Linguistics and the 7th International Joint Conference on Natural Language Processing, vol, 1, Long Papers, pp. 687–696 (2015)

15. Xu, J., Qiu, X., Chen, K., Huang, X.: Knowledge graph representation with jointly structural and textual encoding. In: IJCAI International Joint Conference on Artificial Intelligence, pp. 1318–1324 (2017)

16. Xie, R., Liu, Z., Jia, J., Luan, H., Sun, M.: Representation learning of knowledge graphs with entity descriptions. In: AAAI, pp. 2659–2665 (2016)

17. Vargas, M.R., de Lima, B.S.L.P, Evsukoff, A.G.: Deep learning for stock market prediction from financial news articles. In: IEEE International Conference on Computational Intelligence and Virtual Environments for Measurement Systems and Applications (CIVEMSA), pp. 60–65 (2017)

18. Abadi, M., et al.: TensorFlow: Large-Scale Machine Learning on Heterogeneous Distributed Systems (2016)

19. Chan, W.C.: Stock price reaction to news and no-news: Drift and reversal after headlines. J. Financ. Econ. **70**, 223–260 (2003)

20. Fehrer, R., Feuerriegel, S.: Improving decision analytics with deep learning: The case of financial disclosures, pp. 1–39 (2015)

21. Prosky, J., Song, X., Tan, A., Zhao, M.: Sentiment Predictability for Stocks (2017)

22. Weng, B., Ahmed, M.A., Megahed, F.M.: Stock market one-day ahead movement prediction using disparate data sources. Expert Syst. Appl. **79**, 153–163 (2017)

23. Lin, X., Yang, Z., Song, Y.: Expert systems with applications intelligent stock trading system based on improved technical analysis and echo state network. Expert Syst. Appl. **38**, 11347–11354 (2011)

24. Hoffmann, P.: Recognizing contextual polarity in phrase-level sentiment analysis. In: Proceedings of the conference on human language technology and empirical methods in natural language processing, pp. 347–354. Association for Computational Linguistics. pp. 347–354 (2005)

25. Mikolov, T., Chen, K., Corrado, G., Dean, J.: Efficient Estimation of Word Representations in Vector Space, pp. 1–12 (2013)

26. Kim, Y.: Convolutional Neural Networks for Sentence Classification (2014)

27. Kraus, M., Feuerriegel, S.: Decision support from financial disclosures with deep neural networks and transfer learning. Decis. Support Syst. **104**, 38–48 (2017)

28. LeCun, Y., Bengio, Y., Hinton, G.: Deep learning. Nature **521**, 436–444 (2015)

29. Nikfarjam, A., Emadzadeh, E., Muthaiyah, S.: Text mining approaches for stock market prediction. In: The 2nd International Conference on Computer and Automation Engineering (ICCAE), vol. 4, pp. 256–260 (2010)

30. Sokolova, M., Japkowicz, N., Szpakowicz, S.: Beyond accuracy, F-Score and ROC: a family of discriminant measures for performance evaluation. In: Sattar, A., Kang, B.-h. (eds.) AI 2006. LNCS (LNAI), vol. 4304, pp. 1015–1021. Springer, Heidelberg (2006). https://doi.org/10.1007/11941439_114

Sample Dropout for Audio Scene Classification Using Multi-scale Dense Connected Convolutional Neural Network

Dawei Feng[1], Kele Xu[1,2(✉)], Haibo Mi[1], Feifan Liao[2], and Yan Zhou[2]

[1] Science and Technology on Parallel and Distributed Laboratory,
School of Computer, National University of Defense Technology,
Changsha 410073, China
davyfeng.c@gmail.com, kelele.xu@gmail.com, haibo_mihb@126.com
[2] School of Information and Communication,
National University of Defense Technology, Wuhan 430010, China

Abstract. Acoustic scene classification is an intricate problem for a machine. As an emerging field of research, deep Convolutional Neural Networks (CNN) achieve convincing results. In this paper, we explore the use of multi-scale Dense connected convolutional neural network (DenseNet) for the classification task, with the goal to improve the classification performance as multi-scale features can be extracted from the time-frequency representation of the audio signal. On the other hand, most of previous CNN-based audio scene classification approaches aim to improve the classification accuracy, by employing different regularization techniques, such as the dropout of hidden units and data augmentation, to reduce overfitting. It is widely known that outliers in the training set have a high negative influence on the trained model, and culling the outliers may improve the classification performance, while it is often under-explored in previous studies. In this paper, inspired by the silence removal in the speech signal processing, a novel sample dropout approach is proposed, which aims to remove outliers in the training dataset. Using the DCASE 2017 audio scene classification datasets, the experimental results demonstrates the proposed multi-scale DenseNet providing a superior performance than the traditional single-scale DenseNet, while the sample dropout method can further improve the classification robustness of multi-scale DenseNet.

Keywords: Sample dropout · Audio scene classification
Convolutional neural network · Multi-scale

1 Introduction

Acoustic scene classification (ASC) aims to distinguish between the various acoustical scenes, and effective identification of scenes through the analysis of unstructured patterns has many potential applications, such as, intelligent monitoring systems, context aware devices design and so on. Multiple overlapping

© Springer Nature Switzerland AG 2018
K. Yoshida and M. Lee (Eds.): PKAW 2018, LNAI 11016, pp. 114–123, 2018.
https://doi.org/10.1007/978-3-319-97289-3_9

sound sources are contained in the acoustic mark of a certain scene, as a result, ASC is of a great challenge despite the sustainable efforts have been made.

From the perspective of scene classification, different methods have been tested in the computer vision field, and dramatic progress has been made during last two decades, especially with the improvements of local invariant feature, such as (SIFT [1], SURF [2]) and convolutional neural network [3]. Nevertheless, compared to the image-based scene classification, audio-based approach is still under-explored. The state-of-the-art scene-classification methods using audio are not able to provide comparable accuracy with comparison to the image-based methods [4]. However, the audio is more descriptive and salient than the images in some practical situations.

In the past several years, an increasing interest has been observed, which aims to find more robust and efficient approaches for acoustic scene classification and sound event detection, by using both supervised learning and unsupervised-learning methods. Specifically, the first Detection and Classification of Acoustic Scenes and Events (DCASE) 2013 challenge [5] was organized by the IEEE Audio and Acoustic Signal Processing (AASP) Technical Committee, aims to solve the problem of lacking common benchmarking datasets, and has stimulated the research community to explore more efficient methods. Since the release of the relatively larger labeled data, there has been a plethora of efforts made for the audio scene classification task [6–8].

Recently, deep learning have achieved convincing performance in different fields, ranging from computer vision [3], speech recognition [9] to natural language processing [10]. Extensive deep learning architectures have been explored for the audio signal processing, for example, auto-encoder, convolutional neural network, recurrent neural network, and different regularization methods are also tested for the task.

Most of the previous attempts aimed to apply the deep learning by modifying the CNN architectures. In this paper, we aim to improve the ASC performance by using the multi-scale DenseNet and culling sample-based regularization method. In more detail, multi-scale DenseNet is employed to extract multi-scale information embedded in the time-frequency of the audio signal. Moreover, unlike previous attempts to dropout hidden layers in the neural network training, we explore the low-variance sample dropout approaches, with the goal to culling the "outliers" in the training samples. After removing the specified samples in the training data set, the neural network classifier is trained with the remaining examples to obtain robust models. Using the DCASE 2017 audio scene classification dataset [8], our experimental evaluation shows that the proposed method can improve the robustness of the classifier.

The paper is organized as follows: Sect. 2 gives a short summary for the related work. Section 3 presents the data used and the experimental setup, while Sect. 4 describes the multi-scale DenseNet. The approach to cull samples in the training set is given in Sect. 5. The experimental results are presented in Sect. 6, while Sect. 7 concludes this paper.

2   Related Work

ASC is a complicated issue which aims at distinguishing acoustic environments solely depended on the audio recordings of the scene. During last decades, various feature extraction algorithms (representing the audio scenes), and classification models have been proposed in previous works. The most popular baseline is Gaussian Mixture Model (GMM) or Hidden Markov Model (HMM), by using the Mel-Frequency Cepstral Coefficients (MFCCs) [7]. Shallow-architecture classifier, such as, Support Vector Machines (SVM) [11] and Gradient Booting Trees (GBM) [12], were also tested for the classification task. Moreover, non-negative matrix factorization (NMF) approach can be utilized to extract subspace representation prior to the classification.

Recently, many works demonstrated that deep neural network can improve the classification accuracy while no handcrafted features are needed. In brief, for the ASC task, the main modifications of deep learning-based approaches can be divided into three categories: deep learning using different representations of the audio signal [6,13]; more sophisticated deep learning architectures classifiers [14,15] and the applications of different regularizations methods to train the deep neural network [16].

Compared to the traditional methods, which commonly involved training a Gaussian Mixture Model (GMM) on the frame-level features [7], deep learning-based methods can achieve better performance [8]. The well-known deep neural network models include deep belief network (DBN) and auto-encoders, convolutional neural network (CNN), recurrent neural network (RNN). Here, CNN is selected as the classifier due to its high potential to identify the various patterns of audio signals. Moreover, unlike previous attempts, we explore the use of multi-scale DenseNet as additional features may be embedded in different time range. The combination of features from multi-scale may lead to more salient feature representations for the classification task.

On the other hand, the neural networks are vulnerable to the outliers in the training set, and the outliers in the training set have a high negative influence on the trained model. As a result, the pre-processing of training samples is a key factor for the audio scene classification accuracy [17], while it is often under-explored in previous studies. In this work, we explore to cull the train samples of low-variance, which can be viewed as the noise (or outliers), with the goal to improve the accuracy.

3   Audio Scene Classification Datasets and Experimental Setup

As aforementioned, for the audio scene classification task, the well-collected data sets include: DCASE 2013 dataset [5], DCASE 2016 dataset [7], DCASE 2017 dataset [8], Rouen dataset [18] and ESL dataset [19]. We evaluate the proposed method on the TUT audio scene classification 2017 database [8], as this data is of large scale and covers 15 different acoustic scenes.

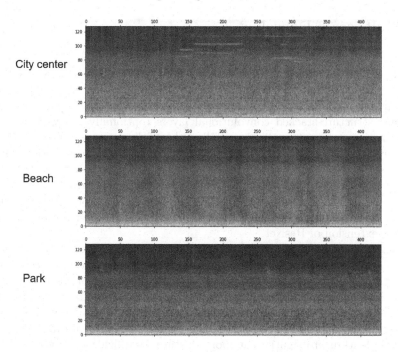

Fig. 1. Mel filter bank energy of different audio scene.

In more detail, the database consists of stereo recordings, which were collected using 44.1 kHz sampling rate and 24-bit resolution. The recordings came from various acoustic scenes, which have distinct recording locations. 3–5 min long audio was recorded for each sample, all of samples are divided into four cross-validation folds. And the audios were split into 10-s segments. The acoustic scene classes considered in this task were: bus, cafe/restaurant, car, city center, forest path, grocery store, home, lakeside beach, library, metro station, office, residential area, train, tram, and urban park.

As the input of the different deep neural network architectures, it can be either the raw audio signal or the time-frequency representation of the raw audio. Presently, most of the audio-related classification and detection system relied on the hand-crafted time-frequency representations of the audio signal. For example, Mel-frequency cepstrum (MFCCs) are widely used in speech recognition. However, MFCCs are developed inspired by the human speech production process, which assumes sounds are produced by glottal pulse passing through the vocal tract filter. MFCCs discard useful information about the sound, which restricts its ability for recognition and classification. In recent years, Mel Bank Features have been widely used in speaker recognition. In this paper, Mel-filter bank feature is used for our experiments, as mel filterbanks can provide better performance, and the mel-bank is reweighted to the same height (shown in Fig. 1).

4   Multi-scale DenseNet

Convolutional neural network (CNN) has a great potential to identify the various salient patterns of audio signals. There are numerous variants of CNN architectures in the literature. However, their basic components are very similar. Since the starting with LeNet-5 [20], convolutional neural networks have typically standard structure-stacked convolutional layers (optionally followed by batch normalization and max-pooling) are followed by fully-connected layers [21].

Many recent works claim that deeper CNN can provides better performance, as demonstrated by the progress on the image-classification task by using AlexNet [3], VGGNet [22], ResNet [23] architectures. Unlike traditional sequential network architectures such as AlexNet, ResNet is a form of architecture that relies on network-in-network block. DenseNet [24] is a new architecture and it is a logical extension of ResNet.

In more detail, DenseNet has a fundamental building block, which connects each layer to every other layer in a dense manner. There is a direct connection between any two layers in each dense block, and the input for each layer is the union of the outputs from all the previous layers. Compared to the conventional CNN, DenseNet not only performs better in image classification, but also provide a higher utilization rate for the original data and less feature information loss. It reinforces feature propagation, supports feature re-utilization, solves vanishing gradient problems effectively and significantly reduce the number of parameters.

For the audio scene classification task, the input for the DenseNet can the time-frequency representation of the raw audio signal. In this paper, as aforementioned, the Mel filter-bank energy is selected as the input of neural network. However, it is widely known that the multi-scale information are embedded in the time-frequency representation of the audio signal. Thus, automatically fully extraction of the multi-scale features is of great importance to improve the classification accuracy. A multi-scale DenseNet [25] architecture is given in Fig. 2.

As shown in Fig. 2, the proposed system architecture comprises of dense convolutional blocks with direct connections from any layer to all subsequent layers to improve the information flow on 128×128 input images. Four dense blocks with unequal numbers of layers make up the DenseNet used in our experiments.

A convolution with 64 output channels is performed on the input images in front of the first dense block. For convolutional layers with kernel size 3×3, one-pixel padding is applied at each side of the inputs to keep the feature-map size fixed. The layers between two contiguous dense blocks are referred as transition layers for convolution and pooling, which contain 1×1 convolution and 2×2 average pooling. A bottleneck layer is used by using a 1×1 convolution before each 3×3 convolution in order to reduce the number of input feature-maps and improve computational efficiency. At the end of the last dense block, a global average pooling and a softmax classifier are employed.

The multi-scale dense block is employed to capture the multi-scale features embedded within the Mel filter bank features. The length of the convolution kernel is longer in the upper layer and can generate more feature maps to extract long-term and medium-term features in the Mel filter bank energy representation

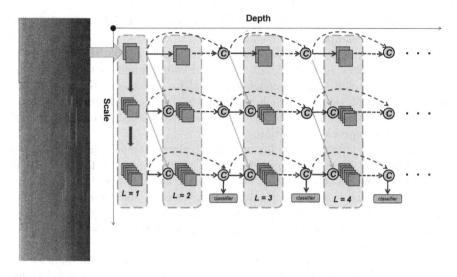

Fig. 2. The multi-scale DenseNet architecture employed for audio scene classification task.

of the audio signal. In the lower layer, the convolution kernel receives all outputs in the previous layers and the original input. As a result, it allows us to decrease the length of the convolution kernel and prompt the feature maps to grasp more short-term features. Since the network structure supports to re-utilize the feature maps, the multi-scale dense block concatenates all periodic features with different lengths into the transition layer for further manipulations. After this block, the model can extract a certain number of multi-scale features.

5   Culling Training Samples for Convolutional Neural Network

It is widely known that convolutional neural networks have a natural tendency towards overfitting, especially when the training data size is small. DCASE 2017 audio scene classification dataset have much larger data size with comparison to the datasets used in the previous studies. However, compared the data size to train a neural network for the image classification task, the size is still small. Dropout and data augmentation are proved to be effective regularization approaches to train the convolutional neural network. In more detail, dropout of neural in the hidden layer is widely used in a plethora of literatures, and the activation of every hidden unit is random randomly removed with a pre-defined probability [26]. While, another common method to reduce overfitting on image data is to artificially enlarge the dataset using label-preserving transformations for data augmentation. However, artificially enlarge the dataset may induce more ambiguous data (which can be considered as the outliers) on which the model has low confidence and thus high loss. In the audio scene classification task, it is widely known that the convolutional neural networks are vulnerable to the ambiguous samples.

In this paper, unlike previous attempts, we aim to reduce the training set by removing outliers. In more detail, we tested two different kinds of methods. First, inspired by the silence remove employed for the speech recognition, we remove the training samples, which associated with low-magnitude of the audio signal segment. It is well-known that silence-remove can boost the performance of the automatic speech recognition.

Secondly, we explore to cull low-variance training samples directly. As can be seen from the Fig. 1, flatten regions exits in the mel-bank features of the audio signal. However, these flatten regions may contain few useful information during the network training as pixel values are almost the same in the whole sample. Our insight is that by removing the low-variance samples, we can make our model to be more robust without changing either the network architecture or training method.

In our experiments, we trained the classifier by using the multi-scale DenseNet to test every example in the training set, and considered examples whose variances were below a threshold as outliers, which will be removed during the majority voting.

6   Experimental Results

In our experimental stage, the average accuracy is calculated by using pre-defined four folds for the cross-validation. For comparison, the baselines are also given in our experiments. The baseline includes a Gaussian mixture model (GMM) (with 16 Gassians per class).

In more detail, the baseline system used here consists of 60 MFCC features and a Gaussian mixture model based classifier. MFCCs were calculated using 40-ms frames with Hamming window and 50% overlap and 40 mel bands. They include the first 20 coefficients (including the 0th order coefficient) and delta and acceleration coefficients calculated using a window length of 9 frames. A GMM model with 32 components was trained for each scene class. To train the multi-scale DenseNet, we used Keras with tensorflow backend, which can fully utilize GPU resource. CUDA and cuDNN were also used to accelerate the system.

The audio scene classification accuracies obtained by different dropout ratio are given in Table 1. During our experiments, the silence remove-based sample dropout method does not boost the performance, thus the culling samples of low-variance, using mel filterbank energy representation, is reported in our experiments. Due to the page limitation, only the multi-scale DenseNet-based audio scene classification accuracy is reported, by using different sample dropout ratio. As can be seen from the table, the sample dropout can improve the performance for the classification task. While, 0.01 % of low-variance samples dropout processing provides superior performance. The dropout ratio 0.01 is employed in our following experiments.

Table 2 provides a further quantitative comparison between different methods. In our experiments, both original DenseNet and multi-scale DenseNet

Table 1. Multi-scale DenseNet-based audio scene classification accuracy using different sample dropout ratio

Sample dropout ratio	Cross-validation	Evaluation
0	80.3%	70.6%
0.01	83.4%	72.5%
0.1	82.5%	71.8%
0.2	80.2%	70.7%

showed better performance against the baseline. Without any data augmentation, the original cross-validation accuracy of multi-scale DenseNet is 80.4%, while the original (single scale) DenseNet provides the 78.3% accuracy. With sample dropout, the performance of original DenseNet and multi-scale DenseNet are boosted. The performance of baseline is also improved with sample dropout.

For the performance on the evaluation dataset, the accuracy of baseline was improved from 61.0% to 63.4% with the sample dropout. The accuracy of original DenseNet was improved from 68.8% to 69.5% with the sample dropout, while the accuracy of multi-scale DenseNet was improved from 68.8% to 69.5% with the sample dropout.

Table 2. Audio scene classification accuracy using single/multi-scale DenseNet and sample dropout (0.01 ratio)

Method	Cross-validation	Evaluation
Baseline (GMM)	74.8%	61.0%
DenseNet	78.3%	68.8%
Multi-scale using DenseNet	80.4%	70.5%
Baseline (GMM) with sample dropout	76.2%	63.4%
DenseNet with sample dropout	80.6%	69.5%
Multi-scale using DenseNet with sample dropout	83.4%	72.5%

As can be seen from the table, the results demonstrates that multi-scale DenseNet showing a better performance than the original single-scale DenseNet. It may imply that additional features can be extracted from multi-scale, which can improve the accuracy for the audio scene classification task. Moreover, the sample dropout method can boost the performance further, which demonstrates the effectiveness of proposed method.

7 Conclusion

In this paper, we have presented the multi-scale DenseNet-based method for the multi-class acoustic scene classification. To summarize, the contributions of this paper are twofold: firstly, we explore a multi-scale CNN architecture for the classification task. To the best knowledge of the authors, this is the first attempt to employ multi-scale DenseNet for the audio scene classification task. Secondly, we propose a novel sample dropout method, and experiments demonstrate that by employing the proposed sample dropout approach, the classification performance can be improved further. For future work, we will conduct a quantitative comparison between different widely-used CNN architectures, which can be helpful to design specified architecture for the audio scene classification task. Moreover, the transfer learning-based audio scene classification also needs to be explored in future work.

Acknowledgement. This study was supported by the Strategic Priority Research Programme (17-ZLXD-XX-02-06-02-08).

References

1. Lowe, D.G.: Distinctive image features from scale-invariant keypoints. Int. J. Comput. Vis. **60**(2), 91–110 (2004)
2. Bay, H., Tuytelaars, T., Van Gool, L.: SURF: speeded up robust features. In: Leonardis, A., Bischof, H., Pinz, A. (eds.) ECCV 2006. LNCS, vol. 3951, pp. 404–417. Springer, Heidelberg (2006). https://doi.org/10.1007/11744023_32
3. Krizhevsky, A., Sutskever, I., Hinton, G.E.: Imagenet classification with deep convolutional neural networks. In: Advances in Neural Information Processing Systems, pp. 1097–1105 (2012)
4. Dixit, M., Chen, S., Gao, D., Rasiwasia, N., Vasconcelos, N.: Scene classification with semantic fisher vectors. In: 2015 IEEE Conference on Computer Vision and Pattern Recognition (CVPR), pp. 2974–2983. IEEE (2015)
5. Stowell, D., Giannoulis, D., Benetos, E., Lagrange, M., Plumbley, M.D.: Detection and classification of acoustic scenes and events. IEEE Trans. Multimedia **17**(10), 1733–1746 (2015)
6. Eghbal-Zadeh, H., Lehner, B., Dorfer, M., Widmer, G.: CP-JKU submissions for DCASE-2016: a hybrid approach using binaural i-vectors and deep convolutional neural networks. In: IEEE AASP Challenge on Detection and Classification of Acoustic Scenes and Events (DCASE) (2016)
7. Mesaros, A., Heittola, T., Virtanen, T.: TUT database for acoustic scene classification and sound event detection. In: 2016 24th European Signal Processing Conference (EUSIPCO), pp. 1128–1132. IEEE (2016)
8. Mesaros, A., et al.: DCASE 2017 challenge setup: tasks, datasets and baseline system. In: DCASE 2017-Workshop on Detection and Classification of Acoustic Scenes and Events (2017)
9. Hinton, G., et al.: Deep neural networks for acoustic modeling in speech recognition: the shared views of four research groups. IEEE Sig. Process. Mag. **29**(6), 82–97 (2012)
10. Manning, C., Surdeanu, M., Bauer, J., Finkel, J., Bethard, S., McClosky, D.: The stanford CoreNLP natural language processing toolkit. In: Proceedings of 52nd Annual Meeting of the Association for Computational Linguistics: System Demonstrations, pp. 55–60 (2014)

11. Geiger, J.T., Schuller, B., Rigoll, G.: Large-scale audio feature extraction and SVM for acoustic scene classification. In: 2013 IEEE Workshop on Applications of Signal Processing to Audio and Acoustics (WASPAA), pp. 1–4. IEEE (2013)
12. Fonseca, E., Gong, R., Bogdanov, D., Slizovskaia, O., Gómez Gutiérrez, E., Serra, X.: Acoustic scene classification by ensembling gradient boosting machine and convolutional neural networks. In: Virtanen, T., et al. (eds.) Detection and Classification of Acoustic Scenes and Events 2017 Workshop (DCASE2017), 16 November 2017, Munich, Germany. Tampere (Finland): Tampere University of Technology, pp. 37–41. Tampere University of Technology (2017)
13. Aytar, Y., Vondrick, C., Torralba, A.: Soundnet: learning sound representations from unlabeled video. In: Advances in Neural Information Processing Systems, pp. 892–900 (2016)
14. Marchi, E., Tonelli, D., Xu, X., Ringeval, F., Deng, J., Schuller, B.: The up system for the 2016 DCASE challenge using deep recurrent neural network and multiscale kernel subspace learning. In: Detection and Classification of Acoustic Scenes and Events (2016)
15. Bae, S.H., Choi, I., Kim, N.S.: Acoustic scene classification using parallel combination of LSTM and CNN. In: Proceedings of the Detection and Classification of Acoustic Scenes and Events 2016 Workshop (DCASE2016), pp. 11–15 (2016)
16. Phan, H., Koch, P., Hertel, L., Maass, M., Mazur, R., Mertins, A.: CNN-LTE: a class of 1-x pooling convolutional neural networks on label tree embeddings for audio scene classification. In: 2017 IEEE International Conference on Acoustics, Speech and Signal Processing (ICASSP), pp. 136–140. IEEE (2017)
17. Xu, K., et al.: Mixup-based acoustic scene classification using multi-channel convolutional neural network. arXiv preprint arXiv:1805.07319 (2018)
18. Rakotomamonjy, A., Gasso, G.: Histogram of gradients of time-frequency representations for audio scene classification. IEEE/ACM Trans. Audio Speech Lang. Process. (TASLP) 23(1), 142–153 (2015)
19. Piczak, K.J.: ESC: dataset for environmental sound classification. In: Proceedings of the 23rd ACM International Conference on Multimedia, pp. 1015–1018. ACM (2015)
20. LeCun, Y., et al.: Learning algorithms for classification: a comparison on handwritten digit recognition. Neural Netw. Stat. Mech. Perspect. 261, 276 (1995)
21. Li, B., Xu, K., Cui, X., Wang, Y., Ai, X., Wang, Y.: Multi-scale DenseNet-based electricity theft detection. arXiv preprint arXiv:1805.09591 (2018)
22. Simonyan, K., Zisserman, A.: Very deep convolutional networks for large-scale image recognition. arXiv preprint arXiv:1409.1556 (2014)
23. He, K., Zhang, X., Ren, S., Sun, J.: Deep residual learning for image recognition. In: Proceedings of the IEEE Conference on Computer Vision and Pattern Recognition, pp. 770–778 (2016)
24. Huang, G., Liu, Z., Weinberger, K.Q., van der Maaten, L.: Densely connected convolutional networks. In: Proceedings of the IEEE Conference on Computer Vision and Pattern Recognition, vol. 1, no. 2, p. 3 (2017)
25. Huang, G., Chen, D., Li, T., Wu, F., van der Maaten, L., Weinberger, K.Q.: Multi-scale dense convolutional networks for efficient prediction. arXiv preprint arXiv:1703.09844 (2017)
26. Srivastava, N., Hinton, G., Krizhevsky, A., Sutskever, I., Salakhutdinov, R.: Dropout: a simple way to prevent neural networks from overfitting. J. Mach. Learn. Res. 15(1), 1929–1958 (2014)

LOUGA: Learning Planning Operators Using Genetic Algorithms

Jiří Kučera and Roman Barták$^{(\boxtimes)}$ ⓘ

Faculty of Mathematics and Physics, Charles University,
Malostranské nám. 25, Praha, Czech Republic
bartak@ktiml.mff.cuni.cz

Abstract. Planning domain models are critical input to current automated planners. These models provide description of planning operators that formalize how an agent can change the state of the world. It is not easy to obtain accurate description of planning operators, namely to ensure that all preconditions and effects are properly specified. Therefore automated techniques to learn them are important for domain modelling.

In this paper, we propose a novel method for learning planning operators (action schemata) from example plans. This method, called LOUGA (Learning Operators Using Genetic Algorithms), uses a genetic algorithm to learn action effects and an ad-hoc algorithm to learn action preconditions. We show experimentally that LOUGA is more accurate and faster than the ARMS system, currently the only technique for solving the same type of problem.

Keywords: Planning · Learning · Action models · Genetic algorithms

1 Introduction

Automated planning deals with the problem of finding a sequence of actions that transfer the world from the current state to a desired state. It is a model-based method, where the model formally describes how the actions are changing states of the world. Hence an important aspect of automated planning is obtaining a proper model of actions. In this paper we deal with classical (STRIPS) planning where actions are defined via preconditions and postconditions (effects), each being a set of predicates. The problem that we are addressing in the paper is how to learn these sets of preconditions and postconditions automatically from examples of plans.

There exist various approaches to acquisition of planning domain models. The early works such as EXPO [3] or later works such as STRIPS-TraceLearn [9] improve action models incrementally after observing some problem during plan execution. Another approach learns from expert traces and subsequent simulations [10]. Frequently, the acquisition problem consists of finding the domain model from examples of plans, which is also the topic of this paper. In other words, the problem is to learn a correct state transition function according to

© Springer Nature Switzerland AG 2018
K. Yoshida and M. Lee (Eds.): PKAW 2018, LNAI 11016, pp. 124–138, 2018.
https://doi.org/10.1007/978-3-319-97289-3_10

observed sequences of actions and states. The system ARMS [11] uses partially specified plans as its input, namely each plan consists of the initial state, a sequence actions, and goal predicates. Intermediate states might also be partially specified. Using MAX-SAT, ARMS learns the preconditions and effects of actions. The follower of ARMS called AMAN [12] allows some actions in plans to be wrongly recognized. LOCM [2] and LOCM2 [1] do not use a predicate model of world states but they rather learn finite-state automata for objects in the world. These automata describe how properties of objects are being changed by actions. Similarly, Opmaker2 [6] learns actions as methods to change properties (states) of involved objects and it requires some invariant formulas describing propositions that must be true in any state. ASCoL [5] and LC_M [4] both only extend the LOCM system. ASCoL learns static preconditions for LOCM and LC_M extends it to work with missing and noisy data. Finally LAMP [13] learns more complex action models with quantifiers and logical implications.

We solve the same problem as ARMS, but we relax the condition of knowing the goal predicates. The proposed system LOUGA (Learning Operators Using Genetic Algorithms) learns from valid sequences of actions (not necessarily from plans reaching certain goals as ARMS). We assume that the initial state and a valid sequence of actions is given as input. LOUGA can also exploit partially specified intermediate states and a final state to find more accurate models. We use a classical genetic algorithm to learn the effects of actions and an ad-hoc algorithm to learn the preconditions of actions. We will show experimentally that LOUGA produces more accurate domain models and it is also faster than ARMS.

2   Background and Problem Specification

We work with classical STRIPS planning that deals with sequences of actions transferring the world from a given initial state to a state satisfying certain goal condition. World states are modeled as sets of predicates that are true in those states and actions are changing validity of certain predicates.

Formally, let P be a set of all predicates modeling properties of world states. Then a state $S \subseteq P$ is a set of predicates that are true in that state (every other predicate is false). Each action a is described by four sets of predicates $(B_a^+, B_a^-, A_a^+, A_a^-)$, where $B_a^+, B_a^-, A_a^+, A_a^- \subseteq P, B_a^+ \cap B_a^- = \emptyset, A_a^+ \cap A_a^- = \emptyset$. Sets B_a^+ and B_a^- describe positive and negative preconditions of action a, that is, predicates that must be true and false right before the action a. Action a is applicable to state S iff $B_a^+ \subseteq S \wedge B_a^- \cap S = \emptyset$. Sets A_a^+ and A_a^- describe positive (add list) and negative (del list) effects of action a, that is, predicates that will become true and false in the state right after executing the action a. If an action a is applicable to state S then the state right after the action a will be $\gamma(S, a) = (S \setminus A_a^-) \cup A_a^+$. If an action a is not applicable to state S then $\gamma(S, a)$ is undefined. In this work we use additional assumptions about the applicability of actions, namely $A_a^- \subseteq S$ and $A_a^+ \cap S = \emptyset$. The first assumption says that if an action deletes some predicate from the state then this predicate should be present in the state. Similarly, if an action adds some predicate to the state then the predicate should not be in the state before. These assumptions can be easily included in the action model as $A_a^- \subseteq B_a^+$ and $A_a^+ \subseteq B_a^-$.

In practice, operators are used in the domain model rather than actions. Operator can be seen as a parameterized action. Each operator has a set of attributes and specifies preconditions and effects as predicates over these attributes:

```
(:action move
  :parameters (?o - object ?m - place
               ?l - place)
  :precondition (at ?o ?m)
  :effect (and (at ?o ?l)
               (not (at ?o ?m))))
```

Actions are obtained by substituting constants for the attributes. The planning domain model is then specified by the set of predicates and the set of operators. PDDL modeling language [7] is the most widely used language for modeling planning domains; we will use syntax of that language in our examples.

In our approach, we assume two types of input information. First, there is a partial planning domain model consisting of a set of predicates and a set of operators with attributes but without the description of preconditions and effects. The second type of input is a set of plans, where each plan consists of the initial state and a valid sequence of actions. Partially specified intermediate states or a goal state might also be provided. The information about states can be in three forms: a predicate was observed in the state, a predicate was observed not to be in the state, or the state was fully observed. We do not make any other assumptions about the input data unlike ARMS that presumes that some effect of every action is used by some later action or in the goal state. The task is to complete the domain model by learning preconditions and effects of operators such that the provided input plans are valid plans according to this domain model.

3 LOUGA

The proposed learning approach works in two main stages. First, we will learn action effects using a standard genetic algorithm [8] with some extensions. Second, we will complete the learned action model by learning action preconditions using a polynomial ad-hoc algorithm.

In the following text, we will detail these steps. For the genetic algorithm, we need to encode action effects to a genome. We will show, how to reduce the number of possible genomes by eliminating those violating conditions imposed on action effects. We will also define the fitness function that guides the genetic algorithm and we will show some methods to help the genetic algorithm when being stuck in a local optima. Next, we will show that it is possible to learn the effects predicate by predicate rather than all together. Finally, we will present the method for learning action preconditions.

3.1 Genome Model

Genetic algorithms work with individuals, each individual encoding a solution candidate. In our case, an individual describes effects of operators. First, we generate a list of all operator-predicate pairs such that the operator can use the predicate in its add or delete lists. This property can be easily verified by checking that all attributes of the predicate are among the attributes of the operator. We assume that attributes are typed though this assumption can be relaxed as we will show in the section on experiments. Each operator-predicate pair will be associated with one of three values:

- 0: predicate is not in operator's add and del lists,
- 1: predicate is in operator's add list (positive effect),
- 2: predicate is in operator's delete list (negative effect).

The individual will be the sequence of numbers that corresponds one-to-one to the description of effects of operators.

For example, let us have a model with the following predicates and operators:

```
(:predicates
  (at ?o - (either object briefcase)
      ?l - place)
  (empty ?b - briefcase)
  (free ?o - object)
  (in ?o - object ?b - briefcase)
)
(:action move
  :parameters (?b - briefcase ?m - place
               ?l - place)
)
(:action put-in
  :parameters (?o - object ?p - place
               ?b - briefcase)
)
```

For this model, LOUGA generates the following pairs:

```
1. ((at ?b ?m), (move ?b ?m ?l))
2. ((at ?b ?l), (move ?b ?m ?l))
3. ((empty ?b), (move ?b ?m ?l))
4. ((at ?o ?p), (put-in ?o ?p ?b))
5. ((at ?b ?p), (put-in ?o ?p ?b))
6. ((empty ?b), (put-in ?o ?p ?b))
7. ((free ?o), (put-in ?o ?p ?b))
8. ((in ?o ?b), (put-in ?o ?p ?b))
```

Hence, each individual will be described by a list of length eight. For example, the individual with genome '210 11000' corresponds to the model in which operator (*move ?b ?m ?l*) has predicate (*at ?b ?m*) in its del list and predicate (*at ?b ?l*) in its add list and operator (*put-in ?o ?p ?b*) has predicates (*at ?o ?p*) and (*at ?b ?p*) in its add list.

Note that it is possible to encode operator's preconditions in the same way, but we will present a more efficient method to learn operator's preconditions later.

3.2   Pre-processing

The genome model specifies the search space that the genetic algorithm will explore. We can reduce this space further by eliminating individuals violating constraints of the model. This is done by exploring the example plans and identifying predicates that cannot be present in the add or del lists of specific operators. LOUGA simulates execution of the plan and for each state it finds two sets of predicates: the first set contains predicates that are definitely in the current state and the second set contains predicates that can possibly be in the current state, but it is not certain. Algorithm 1 describes how these sets are constructed and used.

Algorithm 1. Removing possible values of some genes.

Input: plan P; array M representing possible values of genes
Output: modified array M

1: $Q \leftarrow$ predicates from initial state
2: R - empty set of predicates
3: **for all** actions a from P **do**
4:     generate a set of predicates X, which a can use
5:     **for all** $p \in X$ **do**
6:         **if** $p \in Q$ **then**   ▷ p is definitely in current state
7:             M[(a,p),add] = false
8:             $R = R \cup \{p\}$
9:             $Q = Q \setminus \{p\}$
10:        **else if** $p \notin R$ **then**   ▷ p is definitely not in the state
11:            M[(a,p),del] = false
12:            $R = R \cup \{p\}$
13:        **end if**
14:        **if** predicates S were observed after a **then**
15:            $Q = Q \cup S$
16:            $R = R \setminus S$
17:        **end if**
18:        **if** predicates S were observed missing after a **then**
19:            $Q = Q \setminus S$
20:            $R = R \setminus S$
21:        **end if**
22:    **end for**
23: **end for**

Initially, the first set Q contains all predicates from the initial state (line 1) and the second set R is empty (line 2). LOUGA then goes through the actions in the order specified by the plan. For each action it generates the set of all predicates that the action can use. If some predicate is present in the state before the action, LOUGA marks that the action cannot have that predicate in its add list (line 7). If a predicate is definitely not present in that state, LOUGA marks that the action cannot delete it (line 11). All predicates generated for the action are then added to the second set and removed from the first one if they were present in it. If there are some predicates observed in the state after the action, all of these predicates are added to the first set and removed from the second one. After that, LOUGA continues with the next action until it processes the whole plan. The justification of this process is as follows. If some predicate can be modified by the action then that predicate can possibly be part of the next state. If some predicate is in the state and it is not modified by the action then the predicate stays in the state. Also, information about observed predicates can be exploited there (lines 14–21).

For example let us assume this short plan:

```
(:state
  (empty b1)
  (at b1 home)
  (free pencil)
  (at pencil home)
  (at rubber home)
  (free rubber))
(put-in pencil home b1)
(move b1 home office)
```

We know that there are exactly six predicates in the initial state. Action (*put-in pencil home* b1) can work with predicates (*at pencil home*), (*at* b1 *home*), (*empty* b1), (*free pencil*) and (*in pencil* b1). Pairs made of these predicates and operator *put-in* correspond to genes 4–8. Predicates (*at pencil home*), (*at* b1 *home*), (*empty* b1) and (*free pencil*) are definitely present in the state before action *put-in*, which means that the action cannot add them. As a result, genes 4–7 will have disabled value 1 during evolution. Predicate (*in pencil* b1) is not in the initial state, which means that the action cannot delete it and therefore gene 8 will have disabled value 2 during evolution.

After processing the action we move all these predicates to the second set (line 8) and delete them from the first set (line 9), if they are present there. Now the sets contain these predicates:

```
first set Q (definitely in the state)
    (at rubber home)
    (free rubber)
second set R (possibly in the state)
    (empty b1)
    (at b1 home)
    (free pencil)
    (at pencil home)
    (in pencil b1)
```

The next action is (*move b1 home office*). Operator *move* corresponds to genes 1–3, that means that the action can use predicates (*at b1 home*), (*at b1 office*) and (*empty b1*). Predicate (*at b1 office*) is in none of the sets, therefore the action cannot delete it and gene 2 cannot have value 2. Other predicates are already in the second set, so genes 1 and 3 will remain unchanged. We add predicate (*at b1 office*) to the second set and continue with the next action (if any).

3.3   Fitness Function

The genetic algorithm uses a fitness function which evaluates the error rate of the model represented by the individual. We assume three types of errors:

- **add error:** an action tries to add a predicate that is already present in the world state,
- **del error:** an action tries to delete a predicate that is not currently present in the world state,
- **observation error:** a predicate was observed in a state in the original plan, but it is not present in the corresponding state of the plan executed according to the current model, or there is a predicate in a state that should not be present according to observations about the corresponding state in the original plan.

Formally we can define these errors as follows: let S be a state of a plan executed according to the model represented by the individual, T be a set of predicates that were observed in the corresponding state in the input plan, N be a set of predicates that were observed not to be present in the corresponding state, a be the action performed from state S and $p \in A_a^+$, $q \in A_a^-$, $s \in S$ and $t \in T$ be some predicates. Add error occurs when $p \in S$, del error occurs when $q \notin S$, and observation error occurs when $t \notin S$, $s \in N$ or – if T was marked as a fully-observed state – when $s \notin T$.

After all plans are processed, the fitness value of the individual is defined using this formula:

$$(1 - (error_{add} + error_{del})/(total_{add} + total_{del})) * (1 - error_{obs}/total_{obs}),$$

where $error_{add}$, $error_{del}$ and $error_{obs}$ are the numbers of corresponding errors, $total_{add}$ and $total_{del}$ are the numbers of add and delete operations performed in simulation, $total_{obs}$ is the total number of observations about intermediate and goal states plus the number of surplus predicates in fully-observed states. We tried a version of the fitness function that treated all three types of errors identically, but it turned out not to be ideal. When there were too many or too few observations in input data, evolution could get stuck in a local optima that favors good add and delete error rates over the observation error rate or vice versa. Treating observation errors separately solves this problem. We also tried to split the add and delete errors, but that had a marginal effect on efficiency. Obviously, the fitness value 1 means a perfect individual.

3.4   The Genetic Algorithm

LOUGA uses a classical genetic algorithm with one-point crossover and mutation [8]. We tried more sophisticated versions of those operators but we did not find any that would perform significantly better than the standard versions. We extended the standard algorithm by two additional operators applied when the population stagnates for some time (i.e. when the best individual is of certain age).

The first operator is basically 1-step local search starting from the best individual's genome to find all options how to change one gene to get a better individual. Genes are picked one by one and for every gene, every possible value is tried and resulting individuals are evaluated. As every gene has at most three possible values, there are at most $2 * N$ candidate genomes, where N is the length of genome. All individuals that performed better than the current best individual are added to the population.

The second operator is applied when even the local search cannot find a better individual. It stores the best individual and restarts the population. Next time before restarting it tries to use the information from previous runs by crossing the current best individual with the stored genomes from previous runs. If it breaks the stagnation, evolution goes on as before until it starts stagnating again or a perfect individual is found. If the algorithm cannot find a better local optimum even after multiple restarts, the operator deletes the local optima list and the genetic algorithm starts from scratch.

3.5 Learning Effects Predicate by Predicate

In complex[1] domain models, individuals' genomes can be too long for the genetic algorithm to work effectively. LOUGA solves this problem by learning operators' lists separately for each predicate type. It means that an instance of the genetic algorithm is run for each predicate type separately. In each instance, genomes are built only from those operator-predicate pairs that use the correct predicate type and the fitness function ignores observations of predicates of types other than the current predicate type.

This method generates the same genome as if all predicates were learned at once, the learning process is only split into multiple parts. These parts are independent to each other because occurrence of a predicate of one type cannot affect whether occurrence of a predicate of another type is incorrect or not. Therefore this method is sound and yields the same outcome as the standard approach.

3.6 Learning Preconditions

After the add and del sets are learned, the sets of preconditions are generated. LOUGA goes through every plan and for every operator and every relevant predicate it counts the number of cases where the predicate was present before the action was performed and the number of cases where it was not present. After every plan is processed, a positive precondition is created for every such pair that the predicate was always present before the action was performed, and a negative precondition is created for every such pair that the predicate was never present before the action was performed. If evolution gives a perfect individual, this method yields proper precondition lists. Algorithm 2 describes this process formally.

4   Results of Experiments

We evaluated experimentally the contribution of components of LOUGA and we compared LOUGA to ARMS, which is the only other technique solving the same problems. All experiments were run on laptop with Intel Core i5-2410 2.3 GHz processor and 8 GB of RAM. We used five classical domains from planning competitions, namely Blocksworld, Briefcase, FlatTyre, Rover, and Freecell. The Blocksworld domain is a classical planning domain, where a robotic hand builds towers of blocks. In the Briefcase domain, the task is to transport items between locations using a briefcase. The FlatTyre domain deals with the problem of repairing a flat tyre using various tools. The Rover domain models a Mars rover taking pictures and samples. The Freecall domain is an encoding of a card game.

---

[1] As 'complex' models we consider models that have a large number of predicate types and operators. Such models usually have long genomes so the genetic algorithm has to search through a large hypothesis space.

Algorithm 2. Generation of precondition lists

Input: genome G; set of input plans Q
Output: model M

1: create model M represented by G
2: Y, N - integer fields indexed similarly as G; all fields initially 0
3: **for all** $P \in Q$ **do**
4:     $s \leftarrow$ initial state of P
5:     **for all** $a \in P$ **do**
6:         **for all** predicate p, which can be generated by a **do**
7:             $g \leftarrow$ index of gene corresponding to (p,a)
8:             **if** $p \in s$ **then**
9:                 Y[g]++
10:            **else**
11:                N[g]++
12:            **end if**
13:        **end for**
14:        $s \leftarrow s$ after performing a according to M
15:    **end for**
16: **end for**
17: **for all** gene $g \in G$; g corresponds to predicate p and operator o **do**
18:    **if** G[g] = 0 **then**
19:        **if** $N[g] = 0$ && $Y[g] > 0$ **then**
20:            add p to pre_o
21:        **else if** $N[g] > 0$ && $Y[g] = 0$ **then**
22:            add ($not\ p$) to pre_o
23:        **end if**
24:    **else if** G[g] = 1 **then**
25:        add ($not\ p$) to pre_o
26:    **else if** G[g] = 2 **then**
27:        add p to pre_o
28:    **end if**
29: **end for**

The domains were selected because of their various difficulties of how their models can be learned. The easiest ones are the Blocksworld and Briefcase domains, a bit harder is the FlatTyre domain and the hardest ones are the Rover and Freecell domains. Their basic characteristics are given in Table 1.

For each experiment, we randomly generated 200 valid sequences of actions (plans) and performed five-fold cross-validation test by splitting them in five equal parts and running algorithms five times. During each run we used four groups as learning data and the fifth group as a test set. Plans generated for the first three domains had usually 5–8 actions. For domains Rover and Freecell, we generated random walks (without a preset goal) that had about 15–20 actions.

Most tables show runtimes and error rates of generated models (smaller numbers are better). We define errors in similar way as described in the section about fitness function of LOUGA. Add and delete error rates are calculated by dividing the number of errors by the number of performed add or delete actions. Since

Table 1. Comparison of domain models used in experiments.

	Briefcase	Blocksworld	FlatTyre	Rover	Freecell
# object types	3	1	5	7	3
# predicate types	4	5	12	25	11
# operators	3	4	13	9	10
Avg. size of effect lists	3.3	4.5	2	3	5.6
Avg. parameters of operators	2.66	1.5	2.33	4	4.9

ARMS does not work with negative observations, we only evaluate fulfillment of those observations that state which predicates were definitely present in state. Observation error rate is therefore calculated by dividing the number of unfulfilled observations by the total number of predicates observed in intermediate and final states.

The size of population was set to 10, the threshold for local search was set to 7, the threshold for crossover with individuals from previous runs to 10, the threshold for population restart was 15 and mutation probability was 5% with 10% chance for a gene to be switched. From our internal tests we saw that benefits of having bigger population do not outweigh the longer computational time, so we kept the population sizes low. Keeping thresholds high did not provide much benefit neither, because the population did not usually break stagnation in reasonable time anyway.

4.1 Efficiency of Predicate by Predicate Approach

In the first experiment, we will show the effect of splitting the learning problem to multiple smaller problems, where each predicate is learned separately. Both versions of LOUGA reliably find flawless solutions, so we present the runtimes only (Table 2).

Table 2. Performance of LOUGA learning a model predicate by predicate compared to basic version (runtime measured in seconds).

	Pred. by pred.	Basic version
Briefcase	0.22 ± 0.15	0.86 ± 0.64
Blocksworld	0.81 ± 0.66	22.78 ± 40.38
FlatTyre	2.26 ± 0.74	111.76 ± 116.06
Rover	4.11 ± 0.33	$\gg 600$
Freecell	5.73 ± 1.1	$\gg 600$

As expected, the predicate-by-predicate mode performs significantly faster than the basic version. Moreover as the standard deviation indicates, the predicate-by-predicate mode is also more stable. Runtimes of the basic version varied greatly, some runs were even 10 times longer than others. Genetic algorithms usually suffer from such behavior because of the randomness of the method. The predicate-by-predicate mode works more consistently thanks to evolution having a clearer direction. We can say that it generates only one property at a time even though it is composed of many genes. The basic version works with all properties together and improvement in one direction can go hand in hand with step back in other.

4.2 Comparison of GA and Hill Climbing

In the second experiment, we compared the genetic algorithm with the hill climbing approach. In particular, we compared three setups:

- **LOUGA** - the standard version with 10 individuals
- **HC** - hill climbing with 1 individual and local search in every generation; restart when stuck in local optimum
- **GA** - genetic algorithm without local search operator

For the FlatTyre domain, we used inputs with only goal predicates in ending states, so the problem had many solutions. In the other domains, we used plans with complete initial and ending states.

The results (Table 3) show that the genetic algorithm without the local search operator performs much worse than the other two setups. Pure hill climbing performs better on inputs where there are many possible solutions. However if we use complex domains, there is an advantage in incorporating GA, because local search takes a lot of time on big genomes.

Table 3. Runtimes [s] of LOUGA, hill-climbing and a genetic algorithm.

	LOUGA	HC	GA
Briefcase	0.22	0.88	0.36
Blocksworld	0.81	2.22	1.78
Flat tyre (ambiguious)	2.26	1.82	4.2
Rover	4.11	6.92	34.54
Freecell	5.73	11.11	51.52

4.3 Efficiency of Using Types

We assume that objects (constants) are typed, which reduces the number of candidate predicates for preconditions and effects. LOUGA also works with models without types so our next experiment shows the effect of typing on efficiency.

Table 4. Performance of LOUGA on models with and without typing.

	Genome length		Runtime [s]	
	Types	NoTypes	Types	NoTypes
Briefcase	26	32	0.22 ± 0.15	0.97 ± 0.47
Blocksworld	16	108	0.81 ± 0.66	1.91 ± 0.76
FlatTyre	67	405	2.26 ± 0.74	11.47 ± 3.65
Rover	201	2796	4.11 ± 0.33	97.86 ± 7.67
Freecell	291	1481	5.73 ± 1.1	30.7 ± 3.96

The results (Table 4) show that LOUGA can handle domain models without types, thought efficiency decreases significantly. The table also shows the increased size of the genome when types are not used. The added genes (predicates) can be split in two groups. The first group consists of genes that use the unary predicates describing types. These genes do not add any difficulty to the problem, because they are not used in any add or delete lists and thus LOUGA immediately finds a trivial solution for those predicates. The second group consists of genes that use the original predicates. These genes do make the problem noticeably harder. In the Rover domain significantly more of these genes were created, because this domain has more operators and predicate types, and therefore more operator's parameter-predicate's parameter pairs were created by removing typing and more genes needed to be added.

4.4   Comparison to ARMS

Finally, we compared performance of LOUGA and ARMS [11], which is still the most efficient system for this kind of problem. We used two settings there. First, we used example plans with complete initial states, goal predicates, and a small number of predicates observed in intermediate states (every predicate will have 5% chance to be observed). This is the kind of input ARMS was created for. Second, we used plans with complete initial and ending states but no information about intermediate states, which is input that suits LOUGA well. We present the comparison for three domains only where both systems found solutions.

Tables 5 and 6 clearly indicate that LOUGA outperforms ARMS both in terms of runtime and quality of obtained models. From the data we can also see that ARMS has some problems generating delete lists. In many cases, there were zero predicates in delete lists in total. We assume that it is due to ARMS not having enough information about which predicates need to be deleted. LOUGA has less trouble generating those lists thanks to the assumption that a predicate has to be deleted before it can be added again to the world. But in some cases in the first experiment the learned delete lists were not the same as the delete lists of the original model, because the information about what has to be deleted was not sufficiently present in the plans. In the second experiment LOUGA knew that every predicate in the ending states was observed, so it typically found the original models.

Table 5. Comparison of ARMS and LOUGA systems. Inputs with goal predicates and small number of predicates in intermediate states were used.

ARMS	Add ER	Del ER	Pre ER	Obs. ER	Runtime [s]
Briefcase	0.263	–	0.029	0	6.19
Blocksworld	0.409	0.095	0.039	0.001	28.82
Flat tyre	0.319	0.479	0.342	0.003	504.19
LOUGA	Add ER	Del ER	Pre ER	Obs. ER	Runtime [s]
Briefcase	0	0	0	0	0.29
Blocksworld	0	0	0	0	0.64
Flat tyre	0	0	0	0	1.04

Table 6. Comparison of ARMS and LOUGA systems. Inputs with complete goal states were used.

ARMS	Add ER	Del ER	Pre ER	Obs. ER	Runtime [s]
Briefcase	0.318	–	0.032	0	6.72
Blocksworld	0.331	0.061	0.036	0.014	29.64
Flat tyre	0.336	0.507	0.311	0.005	548.09
LOUGA	Add ER	Del ER	Pre ER	Obs. ER	Runtime [s]
Briefcase	0	0	0	0	0.22
Blocksworld	0	0	0	0	0.81
Flat tyre	0	0	0	0	2.26

5 Conclusions

The paper presents a novel approach to learn planning operators from example plans using a genetic algorithm. We presented several techniques to improve performance of the classical genetic algorithm. First, we suggested the preprocessing technique to restrict the set of allowed genomes. Second, we used the genetic algorithm to learn action effects only while the preconditions are learnt separately using an ad-hoc algorithm. Third, we showed that action effects can be learnt predicate by predicate rather than learning the effect completely using a single run of the genetic algorithm. The presented approach LOUGA achieves much better accuracy and it is faster than the state-of-the-art system ARMS solving the same problem.

Acknowledgements. Research is supported by the Czech Science Foundation under the project P103-18-07252S.

References

1. Cresswell, S., Gregory, P.: Generalised domain model acquisition from action traces. In: Proceedings of the 21st International Conference on Automated Planning and Scheduling, ICAPS 2011, Freiburg, Germany, 11–16 June 2011 (2011)
2. Cresswell, S., McCluskey, T.L., West, M.M.: Acquisition of object-centred domain models from planning examples. In: Proceedings of the 19th International Conference on Automated Planning and Scheduling, ICAPS 2009, Thessaloniki, Greece, 19–23 September 2009
3. Gil, Y.: Learning by experimentation: incremental refinement of incomplete planning domains. In: Proceedings of the Eleventh International Conference on Machine Learning (ICML 1994), pp. 87–95 (1994)
4. Gregory, P., Lindsay, A., Porteous, J.: Domain model acquisition with missing information and noisy data. In: Proceedings of the Workshop on Knowledge Engineering for Planning and Scheduling, ICAPS 2017, pp. 69–77 (2017)
5. Jilani, R., Crampton, A., Kitchin, D.E., Vallati, M.: Ascol: A tool for improving automatic planning domain model acquisition. In: Proceedings of AI*IA 2015, Advances in Artificial Intelligence - XIVth International Conference of the Italian Association for Artificial Intelligence, Ferrara, Italy, 23–25 September 2015, pp. 438–451 (2015)
6. McCluskey, T.L., Cresswell, S.N., Richardson, N.E., West, M.M.: Action knowledge acquisition with Opmaker2. In: Filipe, J., Fred, A., Sharp, B. (eds.) ICAART 2009. CCIS, vol. 67, pp. 137–150. Springer, Heidelberg (2010). https://doi.org/10.1007/978-3-642-11819-7_11
7. McDermott, D., et al.: PDDL - the planning domain definition language. CVC TR-98-003/DCS TR-1165, Yale Center for Computational Vision and Control (1998)
8. Mitchell, M.: An Introduction to Genetic Algorithms. MIT Press, Cambridge (1998)
9. Shahaf, D., Chang, A., Amir, E.: Learning partially observable action models: efficient algorithms. In: Proceedings of Twenty-First AAAI Conference on Artificial Intelligence, pp. 920–926 (2006)
10. Wang, X.: Learning by observation and practice: an incremental approach for planning operator acquisition. In: Proceedings of the Twelfth International Conference on Machine Learning (ICML 1995), pp. 549–557 (1995)
11. Yang, Q., Wu, K., Jiang, Y.: Learning action models from plan examples using weighted MAX-SAT. Artif. Intell. 171(2–3), 107–143 (2007)
12. Zhuo, H.H., Kambhampati, S.: Action-model acquisition from noisy plan traces. In: IJCAI 2013, Proceedings of the 23rd International Joint Conference on Artificial Intelligence, Beijing, China, 3–9 August 2013, pp. 2444–2450 (2013)
13. Zhuo, H.H., Yang, Q., Hu, D.H., Li, L.: Learning complex action models with quantifiers and logical implications. Artif. Intell. 174(18), 1540–1569 (2010)

k-NN Based Forecast of Short-Term Foreign Exchange Rates

Haruya Umemoto[1], Tetsuya Toyota[2], and Kouzou Ohara[2(✉)]

[1] Graduate School of Science and Engineering, Aoyama Gakuin University,
Sagamihara, Japan
c5618157@aoyama.jp
[2] Department of Integrated Information Technology, Aoyama Gakuin University,
Sagamihara, Japan
{toyota,ohara}@it.aoyama.ac.jp

Abstract. Recently, day trading, that is, short-term trading that sells/buys financial instruments multiple times within a single trading day is rapidly spreading. But, there are few studies about forecasting short-term foreign exchange rates. Against this background, this work proposes a method of forecasting short-term foreign exchange rates based on k-nearest neighbor (k-NN). The proposed method restricts the search range of k-NN, and avoids forecasting the exchange rate if a certain condition is satisfied. We experimentally evaluate the proposed method by comparing it with an existing k-NN based method, linear regression, and multi-layer perceptron in three metrics: the mean squared forecast error (MSFE), the mean correct forecast direction (MCFD), and the mean forecast trading return (MFTR). The results show the proposed method outperforms the other methods in terms of both MCFD and MFTR, which implies reducing the forecast error does not necessarily contribute to making profits.

Keywords: k-NN · Forecasting exchange rate
Short-term exchange rate

1 Introduction

Nowadays, the foreign exchange market becomes the biggest financial market. To earn huge profits from that market, it is important for a trader to accurately predict the future exchange rates. However, due to its large scale, various factors can exert some influence on the fluctuation of the exchange rate, which makes it difficult to forecast future exchange rates. Meese and Rogoff even report that there are no models that outperform the random walk model in the domain of forecasting exchange rate [5]. Despite of such difficulty, indeed, there exist many attempts to forecast the exchange rate using various models including statistical models, linear models, and non-linear models [6].

© Springer Nature Switzerland AG 2018
K. Yoshida and M. Lee (Eds.): PKAW 2018, LNAI 11016, pp. 139–153, 2018.
https://doi.org/10.1007/978-3-319-97289-3_11

Most of existing studies use long-term rate data in which price data are recorded for each day or each week. On the other hand, recently, the number of people taking the trading method called day trading is increasing. The method generally carries out a trading up to several tens of times per day, so it is known as a low risk and low return method. Since it is difficult to directly apply the techniques for long-term forecasting to day trading, it is crucial for day traders to devise a new method of forecasting short-term exchange rate. However, unfortunately, there exist only few studies that use short-term, or high-frequency exchange rate whose time unit is one minute or shorter [1,8]. There are three major factors that prevent researchers from using the high-frequency data: (1) the difficulty of gathering such high-frequency data, (2) the high computational cost caused by a huge amount of historical data, and (3) the existence of the so-called market micro-structure noise due to the high-frequency.

Under this situation, in this paper, we propose a method of forecasting short-term foreign exchange rates that tries to find and utilize past exchange patterns which are similar to the latest pattern. To this end, the proposed method adopts the k nearest neighbor (k-NN) algorithm. In fact, there exist some studies using k-NN to forecast exchange rates, and one of them shows that using k-NN is effective for forecasting the USD/JPY exchange rate which is known to be difficult to predict [3,7]. In addition to using k-NN to forecast short-term foreign exchange rates, the proposed method further imposes restrictions on its search space to achieve a higher accuracy. More specifically, it only searches historical data within the predefined range. Namely, it does not use too old data for forecasting exchange rates. Besides, the proposed method avoids forecasting the exchange rate if the current fluctuation of rates is extremely high or low. These strategies could alleviate the problems (2) and (3) regarding the difficulty of using short-term data aforementioned.

The proposed method was evaluated on two kinds of short-term (high-frequency) foreign exchange rates in the real world, and compared with an existing k-NN based method, linear regression, and multi-layer perceptron in three metrics: the mean squared forecast error (MSFE), the mean correct forecast direction (MCFD), and the mean forecast trading return (MFTR). Furthermore, we compared the profits earned by these methods through a pseudo-trading conducted under a realistic setting in which the commission was considered. The experimental results showed that the proposed method outperforms the other baseline methods in the two metrics, MCFD and MFTR, and can earn a higher profit than the others do, which suggests that reducing the predictive error (MSFE) does not necessarily lead to earning high profits.

The rest of this paper is organized as follows: Sect. 2 describes the basic notations and briefly explains the existing method based on k-NN, while Sect. 3 gives the details of the proposed method. The experimental settings are provided in detail in Sect. 4, and the experimental results are discussed in Sect. 5. Finally, we conclude this paper in Sect. 6.

2   Forecasting Exchange Rates Using k-NN

In this section, we revisit the k-NN based method for forecasting foreign exchange rates proposed in [3] because we take a similar approach to it. Let E_t be the exchange rate at time t. Then, the problem we consider is to predict the exchange rate at time $t + 1$, E_{t+1}, using the historical data until time t. Indeed, the k-NN based method searches for past fluctuation patterns that are similar to the one that the latest n exchange rates exhibit, and utilizes them to forecast E_{t+1}. We denote the fluctuation pattern consisting of the n latest exchange rates by an n dimensional vector $\boldsymbol{p}_t$ defined as follows:

$$\boldsymbol{p}_t = (E_{t-(n-1)}, E_{t-(n-2)}, \cdots, E_t). \tag{1}$$

Hereafter, we refer to $\boldsymbol{p}_t$ as the target vector. We also define the n dimensional vector $\boldsymbol{q}_{t'}$ $(t' < t)$ in the same fashion for arbitrary n consecutive exchange rates in the data and refer to it as the past vector.

Let $\boldsymbol{q}_{t_1}, \cdots, \boldsymbol{q}_{t_k}$ be the k past vectors found by k-NN. Namely, $\boldsymbol{q}_{t_i}$ is an n dimensional vector $(E_{t_i-(n-1)}, E_{t_i-(n-2)}, \cdots, E_{t_i})$, and the fluctuation pattern it exhibits is assumed to be similar to the one the target vector $\boldsymbol{p}_t$ exhibits. Note that the exchange rate E_{t_i+1} at time $t_i + 1$ is always available in the data D for every $\boldsymbol{q}_{t_i}$ as $t_i < t$. Then, the estimation of E_{t+1}, $\hat{E}_{t+1}$, is defined as follows:

$$\hat{E}_{t+1} = \sum_{i=1}^{k} \frac{E_{t_i+1}}{d_{t_i}} \times \left(\sum_{i=1}^{k} \frac{1}{d_{t_i}}, \right)^{-1} \tag{2}$$

where d_{t_i} stands for the distance between the target vector $\boldsymbol{p}_t$ and the past vector $\boldsymbol{q}_{t_i}$. Equation (2) means that E_{t+1} is estimated as the weighted average of k exchange rates immediately following the k past patterns that are the most similar to the target vector $\boldsymbol{p}_t$. As the distance metric, the dynamic time warping distance (DTW) is used in [3].

3   Proposed Method

The basic idea of the proposed method is improving the accuracy of the afore-mentioned k-NN based method by imposing restrictions on its search strategy. This approach is based on the two observations resulted from applying the k-NN based method to short-term (high-frequency) data. Indeed, we observed the following two tendencies through our preliminary experiments using high-frequency data.

1. Using too old past vectors for the estimation tends to decrease the predictive accuracy.
2. The predictive accuracy tends to decrease when the target vector exhibits an extremely large or small fluctuation.

These observations were obtained only from limited number of data. Thus, a comprehensive analysis is further required to justify the generality of these tendencies. But, in this paper, we assume these hold in general, and refer to them as Hypothesis (1) and (2), respectively.

Assuming these hypotheses, we introduce two strategies into k-NN for high-frequency data to improve its accuracy. First, Hypothesis (1) suggests that it is better not to use too old past data for more accurate estimation. Thus, our first strategy is to restrict the search space of k-NN by introducing a threshold I_{max}. More specifically, we only use the last I_{max} exchange rates to search for the k vectors that are the most similar to the target vector p_t.

Second, Hypothesis (2) suggests that predicting the future exchange rate at the next time step for a target vector that exhibits an extremely large or small fluctuation does not contribute to improving the accuracy. Consequently, as the second strategy, we consider to avoid forecasting the next exchange rate if the fluctuation the target vector p_t exhibits is extremely large or small. To this end, we quantify the degree of the fluctuation of exchange rates during the period corresponding to p_t as follows:

$$M_t = \frac{1}{n} \sum_{i=0}^{n-1} |r_{t-i}|, \tag{3}$$

$$r_t = \log E_t - \log E_{t-1}, \tag{4}$$

where r_t defined by Eq. (4) stands for the logarithmic return at time t. M_t is the average of the absolute value of r_t over n time steps from t to $t - (n-1)$. Then, we introduce the upper-bound and the lower-bound of M_t denoted by C_{max} and C_{min}, respectively, and make a forecast for the exchange rate at time $t + 1$ if and only if $C_{min} < M_t < C_{max}$ holds.

The proposed method forecasts the future exchange rate at time $t + 1$ based on the k vectors that are the most similar to the target vector p_t at time t in the same way as the method in [3]. However, as mentioned above, it only uses the latest I_{max} exchange rates in the past data, and it forecasts the exchange rate if and only if the fluctuation of the target vector falls within the range from C_{max} to C_{min}. It is expected that these restrictions could prevent the predictive accuracy from decreasing, which implies they could contribute to improving the accuracy. The overall process of the proposed method is summarized as follows:

1. Conduct grid search to optimize the hyperparameters n, k, I_{max}, C_{max}, and C_{min} using the latest N exchange rates in the past data as the validation data;
2. Compute M_t for given time t;
3. If $C_{min} < M_t < C_{max}$ holds, then go to Step 4; otherwise terminate;
4. Compose the target vector p_t for time t;
5. Search the latest I_{max} past data for k past vectors that are the most similar to p_t;
6. Output the value of $\hat{E}_{t+1}$ computed by Eq. (2) as the estimation of E_{t+1}.

In accordance with tradition in the domain of financial data analysis, the proposed method converts all of foreign exchange rates to logarithmic returns defined by Eq. (4), and uses the resulting values through the whole process, which means that the estimation $\hat{E}_{t+1}$ is also a logarithmic return. In our experiments, we used $C_{max} = 2C_{min}$ to reduce the computational cost for the grid search, and used the Euclidean distance as the distance metric between vectors.

4 Experimental Settings

4.1 Datasets

In this experiments, we used the historical data of USD/JPY exchange rates recorded minute by minute, which can be obtained from GMO CLICK Securities, inc.[1] Each record of the data consists of four kinds of information with respect to Bid (selling price): opening-price, high-price, low-price, and closing-price. Among them, we only used closing-price in our experiments. As mentioned above, every rate in the data was converted to a logarithmic return using Eq. (4). Indeed, we prepared two datasets that exhibit different trends during their test periods in order to evaluate the robustness of the proposed method. One consists of exchange rates from April 30, 2013, to June 5, 2013, while the other one consists of those from December 7, 2015, to January 15, 2016. For the former dataset, we used 7 days from May 29, 2013 as the period for the test data, while used 7 days from January 8, 2016 for the latter dataset. We refer to the data derived from these 7 days periods as the test data 1 and 2, respectively. Note that the market is closed on Saturday and Sunday, and thus, each period contains 7,200 samples (exchange rates) to be forecasted in total. Figures 1 and 2 respectively show the trends of exchange rates during the test data 1 and 2. From these figures, it is found that the test data 1 shows a downward trend, while it is difficult to find any clear trend that is either downward or upward in the test data 2.

4.2 Baseline Methods

In the experiments, we compared the proposed method with three baseline methods. One is the *k*-NN based method proposed in [3]. Comparing the proposed method with it, we investigate the effect of the restrictions we introduced on the predictive accuracy. The rest two methods are linear regression that is a typical linear model and multi-layer perceptron that is a typical non-linear model. We chose them because they are often used in this domain. To learn the linear regression model, we used the stochastic gradient descent method without regularization for optimization of its hyperparameter. As for the multi-layer perceptron model, we used ReLU (Rectified Linear Unit) as the activation functions of the hidden layer, and the linear function as the activation function of the output layer. As the optimizing algorithm, Adam was adopted. In fact, due to the time limitation, we avoided doing the grid search for the linear regression and the

---

[1] https://www.click-sec.com.

multi-layer perceptron models in the experiments, and determined their hyper-parameters based on the results obtained from the preliminary experiments. More specifically, the number of explanatory variables of the linear regression model was set to 5. Those variables correspond to the latest 5 exchange rates in the past data. The multi-layer perceptron model was given 20 input units that took each of the latest 20 exchange rates in the past data, and 2 hidden layers having 6 units each.

4.3   Evaluation Metrics

To evaluate the performance of the aforementioned methods, we adopted three metrics used in [2].

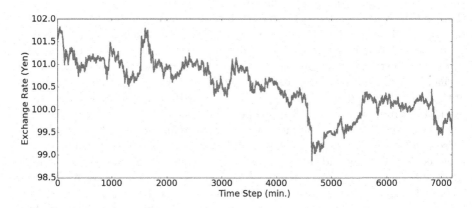

Fig. 1. The fluctuation of the USD/JPY exchange rate in the test data 1

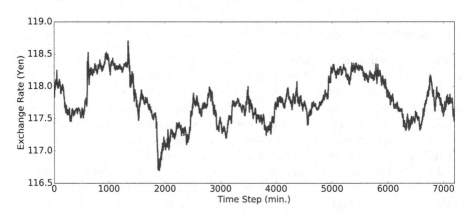

Fig. 2. The fluctuation of the USD/JPY exchange rate in the test data 2

Mean Squared Forecast Error (MSFE) is the metric that evaluates the forecasting error, which is defined as follows:

$$MSFE = \frac{1}{m} \sum_{t=1}^{m} (R_t - \hat{E}_t)^2,$$ (5)

where m is the number of test data, while R_t and $\hat{E}_t$ stand for a real exchange rate and its estimation at time t, respectively.

Mean Correct Forecast Direction (MCFD) is the metric that evaluates the accuracy of the direction of the estimated exchange rate, which is defined as follows:

$$MCFD = \frac{1}{m} \sum_{t=1}^{m} 1(sign(R_t) \times sign(\hat{E}_t) > 0),$$ (6)

where m, R_t, and $\hat{E}_t$ are the same as ones in the definition of MSFE. The function $1()$ returns 1 if the condition $sign(R_t) \times sign(\hat{E}_t) > 0$ is met. As this metric is essentially an error function, the right hand side in the definition in [4] has a coefficient -1. However, we dropped the coefficient in this paper so that one can easily understand the resulting value.

Mean Forecast Trading Return (MFTR) is the metric that evaluates the return itself, which is defined as follows:

$$MFTR = \frac{1}{m} \sum_{t=1}^{m} sign(R_t) \times \hat{E}_t,$$ (7)

where again m, R_t, and $\hat{E}_t$ are the same as ones in the definition of MSFE. Similarly to MCFD, we dropped the coefficient -1 on the right hand side unlike the definition in [4].

It is noted that, in a realistic perspective, [2] points out that it is important to evaluate the performance of predictive models by using MCFD or MFTR that are economical criteria, rather than using MSFE that is a statistical criterion.

4.4   Experimental Method for Exchange Rate Forecast

In our experiments, we determined the training data against each day in the test period. Indeed, the proposed method forecasted exchange rates in a day of the test period using past exchange rates in 30 days immediately before that day. More precisely, we assumed that there exist 22 business days in a month on average, and used $22\,(days) \times 24\,(hours) \times 60\,(minutes) = 31,680$ samples (exchange rates) that were observed immediately before the test day as the training data. Here, a business day means the day when the market is open[2]. It is

---

[2] The second dataset contains the year-end and New Year holidays during which the market is closed. Therefore, we had to use a longer period in the second dataset than in the first one to gather the same size of training data for each test day.

noted that the proposed method searched for only the latest I_{max} exchange rates in the training data rather than the whole of the 31,680 past exchange rates. On the other hand, the k-NN based method used the whole of the past rates before the test day since it has no restriction on its search space. Both of them used the data included in the latest 7 days in the training data as the validation data to optimize their hyperparameters. The hyperparameters to be optimized are n (the dimension of the pattern vector), k (the number of nearest neighbors to be considered), I_{max} (the number of samples to be searched for), and C_{min} (the lower-bound of the degree of the fluctuation M_t) for the proposed method, while n and k for the k-NN method. The parameters were optimized so that they maximize the value of MCFD for the validation data that are independent of the test data. To this end, based on the results of our preliminary analysis, we adopted the values of $\{5, 10, 15, 30\}$ for n, $\{5, 9, 13, 17\}$ for k, $\{1000, 2000, 4000, 10000\}$ for I_{max}, and $\{0.0002, 0.00025, 0.0003, 0.0004\}$ for C_{min} when conducting the grid search. The linear regression and multi-layer perceptron models were built using the same 31,680 past exchange rates as the ones the proposed method used. For fair comparison, all of the methods forecasted the same data, which means that the three baseline methods avoided forecasting an exchange rate if the proposed method did not forecast its value according to Step 3 described in Sect. 3. We did not take into account the canceled samples for calculation of the evaluation metrics.

4.5   Experimental Method for Pseudo-Trading

In addition to the above evaluation, we conducted pseudo-trading and compared the returns earned by the proposed method with the ones obtained by the baseline methods. The detailed settings of the pseudo-trading are described below.

Spread and Commission

In general, the Bid price (selling price) in the currency trading is set to a price that is slightly higher than the Ask price (buying price), and the difference between them is called the spread. The spread is practically considered as the commission in trading, so many currency trading platform companies make the commission free. The spread in our experiments is set to 0.003 Yen per 1 Dollar according to GMO CLICK Securities, inc., and the currency swap is not considered. The spread is usually fixed, but may be changed if an extensive price volatility is observed in the market. In our pseudo-trading, we never changed the spread because the spread data were not available.

Initial Asset and Investment Amount of Each Trading

For better understanding of the fluctuation of the asset, we set the initial asset to 1 Yen, and the investment amount of each trading to 1 Yen no matter whether the asset is higher or lower than 1.

Trading Algorithm

We adopted a trading algorithm that executes buying when the forecasted exchange rate becomes higher than the current rate and otherwise executes selling. The calculation of profits and losses is executed one unit time later, *i.e.*, a minute later, in our trading. But, it is not calculated and the position (buying or selling) is held when the direction of the newly forecasted value is identical to the one of the latest forecasted value. Note that the proposed method may avoid forecasting the next exchange rate as mentioned above. Thus, we considered two strategies for the proposed method in that case: one is settling the profits and losses, and the other is holding the position. We refer to the proposed method that adopts the former strategy as "Proposed k-NN", while the one adopting the latter strategy as "Proposed k-NN Hold". Similarly, we consider the same two strategies for the other methods and add the postfix "Hold" to the name of the method that adopts the latter strategy when avoiding forecasting the next exchange rate.

5 Experimental Results

5.1 Evaluation in Forecasting Exchange Rates

First of all, we investigated the performance of the three baseline methods in the case that they forecast all of the samples in the test data in terms of MSFE, MCFD, and MFTR. Here, the proposed method is not considered because it may cancel some test data. Tables 1 and 2 show the results for the test data 1 and 2, respectively. In these tables, k-NN, LR, and MLP respectively stand for the k-NN based method, the linear regression model, and the multi-layer perceptron model. From these results, we can find that MLP achieves the highest accuracy in MSFE, but the difference between MLP and LR is not so significant. On the other hand, k-NN is the worst in MSFE. This is because MLP and LR try to minimize Mean Squared Error during their learning process unlike k-NN. In MCFD, LR is the best model, while k-NN and MLP are nearly equal to random forecasting. LR achieves the highest accuracy also in MFTR, while k-NN outperforms MLP. The value of MFTR for MLP is less than zero, which means its trading has a deficit. Note that MLP achieved the lowest MSFE, while its MFTR was the worst. These facts suggest that minimizing MSFE (the estimation error) does not necessarily contribute to maximizing MFTR (the profit of tradings). On the other hand, k-NN and MLP are comparable with each other in MCFD, but the value of MFTR for k-NN is positive unlike MLP, which means k-NN was succeeded in making a profit. This difference between k-NN and MLP seems to be attributed to the fact that k-NN achieves a higher accuracy than MLP when the market largely fluctuates.

Next, we compared the proposed method with the three baseline methods in terms of the same metrics. As mentioned above, in this case, the baseline methods avoided forecasting the exchange rates that were not forecasted by the proposed method for fair comparison. In fact, 424 exchange rates out of the test samples were not forecasted due to their degree of fluctuation being larger than the upper-bound C_{max}, and 6,436 rates were not forecasted due to their degree of fluctuation being less than the lower-bound C_{min} for the test data 1. Similarly, for the test data 2, 125 and 6,978 rates were not forecasted based on C_{max} and

C_{max}, respectively. Namely, 340 exchange rates were forecasted by each method for the test data 1, while 97 rates for the test data 2. From these results, it is found that, in most cases, the degree of the fluctuation is likely to be less than the lower-bound C_{min}, leading to the avoidance of forecasting the corresponding exchange rates.

The resulting values in each metric are summarized in Tables 3 and 4, in which PM denotes the proposed method. The values in boldface indicate the best ones in the corresponding metric that are significantly different from the second best at the 1% level in the Wilcoxon signed-rank test. LR and MLP achieve the roughly same accuracy in MSFE as ones shown in Tables 1 and 2. On the other

Table 1. Comparison of the three baseline methods on the test data 1

Model	MSFE	MCFD	MFTR
k-NN	$7.329e^{-8}$	0.501	$1.506e^{-6}$
LR	$6.536e^{-8}$	0.517	$9.602e^{-6}$
MLP	$6.533e^{-8}$	0.500	$-2.916e^{-6}$

Table 2. Comparison of the three baseline methods on the test data 2

Model	MSFE	MCFD	MFTR
k-NN	$4.115e^{-8}$	0.496	$5.274e^{-7}$
LR	$3.677e^{-8}$	0.527	$1.624e^{-5}$
MLP	$3.669e^{-8}$	0.489	$-4.658e^{-7}$

Table 3. Comparison of the proposed method and the three baseline methods on the test data 1

Model	MSFE	MCFD	MFTR
PM	$6.552e^{-8}$	**0.562**	**$7.949e^{-5}$**
k-NN	$6.661e^{-8}$	0.532	$2.582e^{-5}$
LR	$6.537e^{-8}$	0.511	$-5.895e^{-6}$
MLP	$6.533e^{-8}$	0.526	$-3.926e^{-6}$

Table 4. Comparison of the proposed method and the three baseline methods on the test data 2

Model	MSFE	MCFD	MFTR
PM	$3.774e^{-8}$	**0.550**	**$6.238e^{-5}$**
k-NN	$3.808e^{-8}$	0.528	$1.580e^{-5}$
LR	$3.742e^{-8}$	0.473	$-1.299e^{-5}$
MLP	$3.740e^{-8}$	0.495	$-1.388e^{-5}$

hand, it is noteworthy that the accuracy of k-NN is much improved. Actually, it gets closer to the accuracy of LR and MLP. This is thought to be due to that, thanks to the thresholds C_{max} and C_{min}, k-NN could avoid samples that were difficult to forecast. These results show that avoiding forecasting exchange rates when the fluctuation of the latest exchange rates is extremely large or small is very effective for k-NN to reduce the forecasting error. For the proposed method (PM), its value of MSFE is smaller than that for k-NN although it is still slightly larger than the ones for LR and MLP. This means that restricting the search space using the threshold I_{max} in the proposed method is effective for further reducing the estimation error and to improve the accuracy. Furthermore, in the two metrics MCFD and MFTR, the proposed method and k-NN outperform LR and MLP, and the proposed method achieves much better performance than k-NN does. Especially, the differences between PM and k-NN in MCFD and MFTR are statistically significant. From the viewpoint of the computation time for forecasting, the proposed method took around 0.2 second for each forecast on average. This is sufficiently small compared to the time interval of 1 minute in the historical data. The similar tendencies are observed both in the test data 1 and 2. From these results, we can say that the combination of the proposed two restrictions introduced to k-NN seems definitely effective for improving the actual profit in trading.

5.2   Evaluation via Pseudo-Trading

We further evaluated the proposed method through the pseudo-trading. First, we compare the results with and without considering the commission (spread) shown in Figs. 3 and 4, respectively. Both results are the ones obtained on the data set 1. From these results, it is found that LR earned the best profit when not considering the commission, but it lost the most amount of asset when considering the commission. This implies that the profit in each trading is less than the

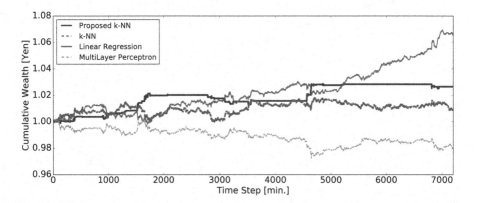

Fig. 3. The results of the pseudo-trading without considering the commission on the test data1

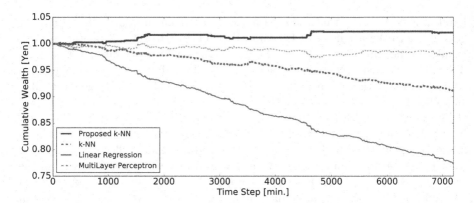

Fig. 4. The results of the pseudo-trading with considering the commission on the test data1

commission when using LR. The same is true of k-NN and MLP. On the other hand, it is found that the proposed method was succeeded in earning the profit both the cases although its profit in the case of not considering the commission is lower than that of LR. Here note that, in this pseudo-trading, the proposed method may avoid forecasting an exchange rate according to Step 3 in Sect. 3, while the other baseline methods forecasted every sample in the test data. To investigate whether avoiding forecasting exchange rates when their fluctuation is extremely large or small is effective on making a profit, or not, we further conducted the pseudo-trading in which all the baseline methods forecasted only the same exchange rates as the ones that were forecasted by the proposed method. The results on the test data 1 and 2 are shown in Figs. 5 and 6. In this case, we considered the commission because it is a more realistic setting. In Fig. 5,

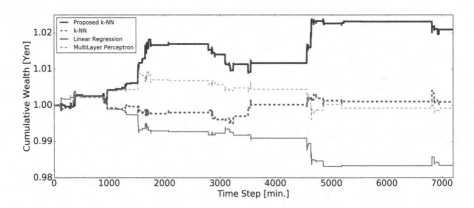

Fig. 5. The results of the pseudo-trading in which all the methods have the chance not to forecast the rate for the test data 1

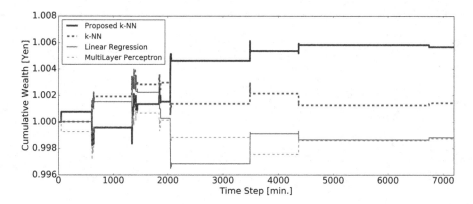

Fig. 6. The results of the pseudo-trading in which all the methods have the chance not to forecast the rate for the test data 2

compared to Fig. 4, it is found that, in the final profit, the proposed method still outperforms the three baseline methods although the deficit of the baseline methods is largely reduced. Especially, the proposed method is much better than *k*-NN both in Figs. 5 and 6. Consequently, we can say again that these results justify restricting the search space of *k*-NN is effective for earning a larger profit.

Finally, we examine the results obtained by taking the strategy of holding the position when each method does not forecast an exchange rate. Figures 7 and 8 show the results on the test data 1 and 2, respectively. Again, in this case, we considered the commission. In Fig. 7, on the one hand, it is found that the final profit of the proposed method is improved compared to Fig. 5. On the other hand, the big advantage of the proposed method over *k*-NN shown in Fig. 5 disappears, and we cannot find any clear difference between them. The similar

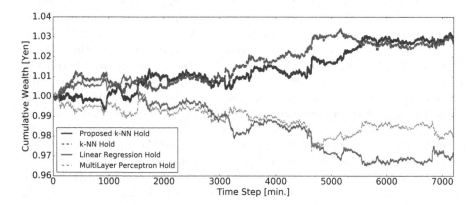

Fig. 7. The results of the pseudo-trading in which all the methods take the strategy of holding the position when avoiding forecasting the rate for the test data 1

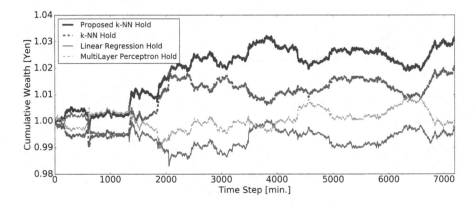

Fig. 8. The results of the pseudo-trading in which all the methods take the strategy of holding the position when avoiding forecasting the rate for the test data 2

tendency can be observed in Figs. 6 and 8, but in this case, the advantage of the proposed method is still found in Fig. 8. As a result, it is concluded that the effect of restricting the search space is effective not only for the proposed method but also the k-NN based method and the advantage brought by the restriction to the proposed method is limited when both the methods take the strategy of holding the position in the case of avoiding forecasting the exchange rate.

6   Conclusion

In this paper, we addressed the problem of forecasting the short-term (high-frequency) foreign exchange rate, and devised a novel method based on k-NN. Based on the observations obtained from our preliminary analysis, the proposed method introduced the two types of restrictions. One is imposed on the search space and excludes the data that are older than the predefined threshold. The other one restricts making a forecast of the exchange rate. More specifically, the proposed method forecast the next exchange rate only if the quantified fluctuation of the rate falls within the range defined in advance. The experimental results on the actual exchange rates showed that these restrictions are effective for reducing the predictive error and improving the final profit and that the proposed method outperforms the existing k-NN based method, the linear regress model, and the multi-layer-perceptron model in terms of the two metrics including the direction of estimated exchange rate and the return. The interesting finding we obtained is that minimizing/reducing the estimation error does not necessarily contribute to maximizing/improving the final profit in forecasting the exchange rate. Besides, through the pseudo-trading, we showed that the proposed method can earn the profit larger than the other methods do in a realistic setting in which the commission is considered.

As immediate future work, we are planning to evaluate the proposed method by using various kinds of data. In addition, it is required to investigate how the performance of the proposed method changes if another distance metric is adopted. DTW (Dynamic Time Warping) distance used in [3] is considered as a possible candidate. But, as its computation is expensive for the high-frequency data, we have to improve the efficiency of the proposed method before introducing this metric. As for the optimization of the hyperparameters, Bayesian optimization is considered as an alternative way. Furthermore, we will extend our method so that it can properly handle a sudden drop of the exchange rate to avoid the risk of lossing the whole of users' bankroll.

References

1. Choudhry, T., McGroarty, F., Peng, K., Wang, S.: High-frequency exchange-rate prediction with an artificial neural network. Intell. Syst. Account. Financ. Manage. **19**(3), 170–178 (2012)
2. Granger, C.W.: Outline of forecast theory using generalized cost functions. Span. Econ. Rev. **1**(2), 161–173 (1999)
3. Kia, A.N., Haratizadeh, S., Zare, H.: Prediction of USD/JPY exchange rate time series directional status by KNN with dynamic time warping AS distance function. Bonfring Int. J. Data Min. **3**(2), 12 (2013)
4. Lee, T.H.: Loss functions in time series forecasting. Int. Encycl. Soc. Sci. **9**, 495–502 (2008)
5. Meese, R.A., Rogoff, K.: Empirical exchange rate models of the seventies: Do they fit out of sample? J. Int. Econ. **14**(1–2), 3–24 (1983)
6. Qian, B., Rasheed, K.: Foreign exchange market prediction with multiple classifiers. J. Forecast. **29**(3), 271–284 (2010)
7. Yao, J., Tan, C.L.: A case study on using neural networks to perform technical forecasting of forex. Neurocomputing **34**(1), 79–98 (2000)
8. Zhou, B.: High-frequency data and volatility in foreign-exchange rates. J. Bus. Econ. Stat. **14**(1), 45–52 (1996)

Multi-dimensional Banded Pattern Mining

Fatimah B. Abdullahi[1] and Frans Coenen[2($\boxtimes$)]

[1] Department of Computer Science, Ahmad Bello University, Zaria, Kaduna, Nigeria
fbabdullahil@yahoo.com
[2] Department of Computer Science, The University of Liverpool, Liverpool, UK
coenen@liverpool.ac.uk

Abstract. Techniques for identifying "banded patterns" in n-Dimensional (n-D) zero-one data, so called Banded Pattern Mining (BPM), are considered. Previous work directed at BPM has been in the context of 2-D data sets; the algorithms typically operated by considering permutations which meant that extension to n-D could not be easily realised. In the work presented in this paper banding is directed at the n-D context. Instead of considering large numbers of permutations the novel approach advocated in this paper is to determine *banding scores* associated with individual indexes in individual dimensions which can then be used to rearrange the indexes to achieve a "best" banding. Two variations of this approach are considered, an approximate approach (which provides for efficiency gains) and an exact approach.

Keywords: Banded patterns in big data · Banded Pattern Mining

1 Introduction

The work presented in this paper is concerned with techniques for identifying "banded patterns" in n-Dimensional (n-D) binary valued data sets, data sets which comprise only ones and zeroes. For ease of understanding, in this paper, the presence of a one is conceptualised as a "dot" (a sphere in 3-D and a hypersphere in n-D, although the term dot will be used regardless of the number of dimensions considered). The presence of a zero is then indicated by the absence of a "dot". The objective is to rearrange the indexes in the individual dimensions so that the dots are arranged along the leading diagonal of the matrix representing the data, or as close to the leading diagonal as possible.

Binary valued data occurs frequently in many real world application domains, examples include bioinformatics (gene mapping and probe mapping) [6,16,27], information retrieval [10] and paleontology (sites and species occurrences) [7,18]. The advantages offered by banding data are fourfold:

1. The resulting banding may be of interest in its own right in that it tells us something about the data; for example it shows us groupings (clusterings) of records that co-occur.

© Springer Nature Switzerland AG 2018
K. Yoshida and M. Lee (Eds.): PKAW 2018, LNAI 11016, pp. 154–169, 2018.
https://doi.org/10.1007/978-3-319-97289-3_12

2. Following on from (1) banding may enhance our interpretability of the data; providing insights that were not clear before.
3. It also allows for the visualisation of the data, which may enhance our understanding of the data.
4. It serves to compress the data, which may consequently enhance the operation of algorithms that work with zero-one data.

The banding of zero-one data in 2-D has a long history; however the idea of Banded Pattern Mining (BPM), as conceived of in this paper, was first proposed in [24,26] (see also [3,20,21]). This early work on BPM was focussed on the heuristically controlled generation and testing of permutations. The generation of permutations is known to be an NP-complete problem, thus the algorithms presented in [24,26] were not easily scalable, hence they were directed at 2-D data. The work presented in this paper is directed at finding bandings in n-Dimensional (n-D) data using the concept of banding scores, an idea first presented in [1,2]. More specifically using a BPM algorithm that uses the banding score concept to assign banding scores to individual indexes in individual dimensions, and reordering the index dimensions accordingly. Two variations of this algorithm are considered, exact and approximate, founded on preliminary work presented in [4,5]. An overall measure of the quality of a banding featured in a given dataset can then be obtained by combining and normalising the individual banding scores to give a Global Banding Score (GBS).

The rest of this paper is structured as follows. Section 2 presents a brief background review of some relevant work. Section 3 then gives the BPM formalism. Section 4 presents the proposed banding score calculation mechanism, while Sect. 5 presents the BPM algorithms that utilise the proposed approximate and exact banding score mechanisms. The evaluation of the proposed approaches is presented in Sect. 6. The paper is concluded in Sect. 7 with a review of the main findings.

2   Related Work

The work presented in this paper is directed at effective banding mechanisms that operate with n-D zero-one data sets. Example applications include: network analysis [19], co-occurrence analysis [18], VLSI chip design [22] and graph drawing [28]. In the case of network analysis the objective is typically community detection. To apply banding the network of interest needs to be represented in the form of an adjacency matrix. By rearranging the rows and columns of the adjacency matrix a banding can be obtained that features groupings of nodes which in turn will be indicative of communities. A specific example can be found in [19] where an American football network data set was used; the communities of interest were teams that frequently played each other. In co-occurrence analysis the aims is the identification of entities that *co-occur*. A specific example can be found in [18] where a paleontological application was considered; here the aim was to match up Neolithic sites with fossil types. The application of banding in the context of VLSI chip design is concerned with the "block alignment problem", where banding allows for the identification of "channels" between the circuit component blocks [22]. In the case of graph drawing we wish to minimise the number of "edge cross overs", this can also be identified using the banding concept [28].

The concept of banded data has its origins in numerical analysis [8] where it has been used in the context of the resolution of linear equations. The banding concept is also related to the domain of reorderable matrixes, reorderable patterns and bandwidth minimisation. Reorderable matrices are concerned with mechanisms for visualising (typically) 2-D tabular data so as to achieve a better understanding of the data [11,12]. The idea of reorderable matrices dates back to the 19th century when Petrie, an English Egyptologist, applied a manual reordering technique to study archaeological data [23,25]. Since then a number of reordering methods have been proposed with respect to a variety of applications. Of note with respect to the work presented in this paper is the BC algorithm [24] which was originally proposed to support graph drawing. The BC algorithm is directed at 2-D data and operates by finding permutations for both rows and columns, such that non-zero entries are as close to each other as possible using a barycentric measure describing the average position of dots in a given column/row. The significance of the BC algorithm with respect to this paper is that it is used as a comparator algorithm with which to evaluate the banding techniques presented.

Reorderable patterns are akin to reorderable matrices, however the idea is to reorder columns and rows so as to reveals some (hidden) pattern of interests [9,17], as opposed to providing a means of facilitating data visualisation. As such the motivation for reorderable patterns can be argued to be the same as that for BPM; the distinction is that the idea of reorderable patterns is concerned with any pre-prescribed pattern P that can be revealed by reordering the columns and rows in a 2-D matrix not just banding (it is also not necessarily directed at zero-one data). This can be viewed as a generalisation of the BPM problem in the sense that the patterns we are looking for in BPM are comprised of dots arranged about the leading diagonal. P in this case would be the locations about the leading diagonal up to a certain distance away, however in the context of the domain of reorderable patterns P can be any shape. The challenge of reorderable patterns is finding a permutation by which the pattern P is revealed.

Bandwidth minimisation [14,15] is concerned with the process of minimizing the bandwidth of the non-zero entries of a sparse 2-D matrix by permuting (reordering) its rows and columns such that the non-zero entries form a narrow "band" that is as close as possible to the leading diagonal. Bandwidth minimisation is clearly also akin to the BPM. The distinction is that bandwidth minimisation is directed at the specific objective of minimising bandwidth to aid further processing of (typically) 2-D matrices, while BPM is directed at data analysis (a by-product of which happens to be bandwidth minimisation and also visualisation).

There has been some limited previous work directed at Banded Pattern Mining (BPM) as conceived of in this paper where BPM is defined as the identification of hidden bandings in zero-one data sets. Of particular note in this context is the work of Gemma et al. [19] who proposed the Minimum Banded Augmentation (MBA) algorithm. The algorithm considers a series of column permutations to produce a number of permuted matrices (ordered matrices). Each column permutation is considered to be fixed whilst row permutations are conducted and evaluated. Two variations of the MBA algorithms have been proposed [19]; the Minimum Banded Augmentation Fixed Permutation (MBA$_{FP}$) algorithm and the Minimum Banded Augmentation Bi-directional Fixed Permutation (MBA$_{BFP}$) algorithm. Both algorithms featured the joint disadvantages of: (i) being computationally expensive; and, as consequence and (ii) of being only applicable to 2-D data. The significance of the MBA$_{FP}$ and MBA$_{BFP}$ algorithms, with respect to the work presented in this paper, is that they were also used to compare the operation of the considered BPM algorithms.

3   BPM Formalism

The data spaces of interest comprise a set of dimensions DIM, where $DIM = \{Dim_1, \ldots, Dim_n\}$. The dimensions are not necessarily of equal size, and each dimension Dim_i comprises a sequence of k index values $\{e_{i_1}, \ldots, e_{i_k}\}$. In 2-D we can conceive of the dimensions as being columns and rows, and in 3-D as columns, rows and slices. Each location may contain zero, one or more dots. The precise distribution of the dots depends on the nature of the application domain. Each dot (hyper-sphere in n-D space) will be represented by a set of coordinates: $\langle c_1, \ldots, c_n \rangle$. The challenge is then to rearrange the indexes in the dimensions so that the dots are arranged along the leading diagonal (or as close to it as possible) taking into consideration that individual locations may hold multiple dots.

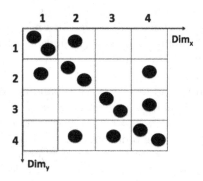

Fig. 1. 2-D multiple dots configuration featuring a banding

4   Calculation of Banding Scores

The fundamental idea underpinning the BPM algorithms considered in this paper is the concept of a banding score [1]. The idea is that given a dimension $Dim_i = \{e_{i_1}, \ldots, e_{i_k}\}$ we can calculate a weighting for each index $e_{i_j} \in Dim_i$ according to the dots that feature at that index. To do this we sum the distances of the dots to the origin of a *modified data space* defined by DIM' where DIM' is the set of dimensions in Dim_i excluding the current dimension to be rearranged. Thus if $DIM = \{Dim_1, Dim_2, Dim_3, Dim_4\}$ and we wish to rearrange the indexes in Dim_2 then $DIM' = \{Dim_1, Dim_3, Dim_4\}$. In the case of a 2-D space, where $DIM = \{Dim_x, Dim_y\}$, this would mean calculating the banding sores for Dim_x with respect to Dim_y, and for Dim_y with respect to Dim_x, which would mean simply summing the y (x) index values and normalising by the maximum indexes taking into account the potential for multiple dots. Given a 2-D space $DIM = \{Dim_i, Dim_j\}$ the banding score for index p in dimension i, bs_{i_p}, will be calculated as shown in Eq. 1.

$$bs_{i_p} = \frac{\sum_{i=1}^{i=|Index|} index_i \in Index \times m_i \in M}{\sum_{i=1}^{i=|Index|}(k_j - i + 1) \times m_i \in M'} \tag{1}$$

where: (i) $Index = \{index_1, index_2, \ldots, index_k\}$ is a list of Dim_j indexes for dots that feature index p in Dim_i; (ii) $M = \{m_1, m_2, \ldots, m_k\}$ is a list of the number of dots at each location corresponding to list I; (iii) M' is a list similar to M but arranged in descending order according to the number of dots at each location; and (iv) k_j is the size of Dim_j. Referring to the example given in Fig. 1 the list I for the first column will be $\{1, 2\}$ (note that the index numbering starts from 1, not 0), the associated list M will be $\{2, 1\}$ and the list M' will also be $\{2, 1\}$. Thus:

$$bs_{x_1} = \frac{(1 \times 2) + (2 \times 1)}{(4 \times 2) + (3 \times 1)} = \frac{4}{11} = 0.363$$

Following on from this $bs_{x_2} = 0.692$, $bs_{x_3} = 0.909$ and $bs_{x_4} = 1.000$. The same scores would be obtained for the y dimension in Fig. 1 because the banding is symmetrical about the leading diagonal. The idea is then to reveal a "best" banding by arranging the indexes, in ascending order from the origin, according to their associated banding scores.

Translating the above to address n-D data, banding scores would be calculated as shown in Eq. 2 where: (i) C a list of locations in the modified data space (data space of size $n - 1$ where $n = |DIM|$); (ii) Max is a list of maximum distances, arranged in descending order such that $|Max| = |C|$; and (iii) M and M' are defined in the same manner as before. The function $dist()$ returns a distance to the origin of the modified data space from the location of its argument expressed as a set of coordinates of the form $\langle c_1, c_2, \ldots, c_{n-1} \rangle$ $(n - 1$ because the modified data space has one dimension less than the original data space). Distance can be determined in a number of manners but two obvious alternatives are Euclidean distance and Manhattan distance. The derivation of the set

Max is not as straightforward as it first seems and is therefore discussed in further detail in Subsect. 5.1. Using Eq. 2, banding scores can be calculated for each dimension and used to iteratively rearrange the indexes in the individual dimensions to reveal a banding.

$$bs_{i_p} = \frac{\sum_{i=1}^{i=|C|} dist(c_i \in C) \times m_i \in M}{\sum_{i=1}^{i=|Max|} (max_i \in Max) \times m_i \in M'} \tag{2}$$

The banding score concept can also be used to calculate a Global Banding Score (GBS) for an entire banding configuration using Eq. 3 where GBS_i is the GBS for dimension i (Dim_i).

$$GBS = \frac{\sum_{i=1}^{i=|DIM|} GBS_i}{|DIM|} \tag{3}$$

The value for GBS_i is then calculated using Eq. 4. Note that each banding is weighted according to its index location as we wish the diagonal around which our dots are arranged to be from the origin of the data space of interest. This means we have to normalise using $\frac{k_i(k_i+1)}{2}$.

$$GBS_i = \frac{\sum_{p=1}^{p=k_i} bs_{i_p} \times (k_i - p + 1)}{\frac{k_i(k_i+1)}{2}} \tag{4}$$

Thus, returning to the configuration given in Fig. 1, using Eq. 4, the value for GBS_x will be calculated as follow:

$$\frac{(0.363 \times 4) + (0.692 \times 3) + (0.909 \times 2) + (1.000 \times 1)}{\frac{4(5)}{2}} =$$

$$\frac{1.452 + 2.076 + 1.812 + 1.000}{10} = 0.634$$

The configuration is symmetrical about the leading diagonal, thus GBS_y will also equal 0.634. The GBS for the entire configuration will then, using Eq. 3, be:

$$\frac{0.634 + 0.634}{2} = 0.634$$

Note that, with reference to Eq. 4, if every cell in a given data space holds exactly the same number of one or more dots, thus no banding at all, the GBS_i for each dimension i will be 1 and the GBS will also be 1. Thus we wish to minimise the GBS value for a configuration to arrive at a best banding. (If the data space contains no dots at all GBS will be 0.).

5   Banded Pattern Mining

The banding score concept, as described above, can be incorporated in BPM algorithms in various ways. Two are presented in this section, Approximate BPM (ABPM) and Exact BPM (EBPM). The first, as the name suggests, produces an approximate (but arguably acceptable) banding while the second, again as the name suggests, produces an exact banding; the advantage of the first is that it is more efficient. At a high level both algorithms work in a similar manner as shown in Algorithm 1. The inputs (lines 1 to 2) are: (i) a dot data set D, comprising a set of tuples of the form $\langle c_1, c_2, \ldots \rangle$, describing the location of each dot in the data space; and (ii) the set of dimensions $DIM = \{Dim_1, Dim_2, \ldots, Dim_n\}$ associated with D. The output is a rearranged data set D that minimises the GBS value. The algorithm iteratively loops over the data space. On each iteration the algorithm rearranges the indexes in the set of dimensions DIM, using the banding score concept, to produce a revised set of dimensions DIM' (line 6). This revised set of dimensions is then used to rearrange D to give D' (line 7). A new GBS is then calculated (using Eqs. 3 and 4). Then, if the new GBS (GBS_{new}) is worse than the current GBS (GBS_{sofar}). The algorithm exits with the previously stored configuration and GBS. Otherwise D, DIM and the value for GBS_{sofar} are updated (lines 12 to 14) and the algorithm repeats.

The ABPM variation of the BPM algorithm considers pairings of dimensions, calculating banding scores using Eq. 1. The advantage, over the EBPM variation, is that the banding score calculation is much more efficient than when calculated using Eq. 2. For the ABPM variation, lines 6 to 8 in Algorithm 1 are replaced with the pseudo code given in Algorithm 2. With reference to Algorithm 2, ABPM operates by considering all possible dimension pairings ij. For each

---

Algorithm 1. Generic BPM Algorithm

1: **Input:** D = Zero-one data matrix subscribing to DIM,
2:        $DIM = \{Dim_1, Dim_2, \ldots, Dim_n\}$
3: **Output:** Rearranged data space D that minimise GBS
4: $GBS_{sofar} = 1.0$
5: **loop**
6:    DIM' = The set of dimensions Dim rearranged using either approximate or exact BPM
7:    D' = The data set D rearranged according to DIM'
8:    GBS_{new} = The GBS for D' calculated using Equations 3 and 4
9:    **if** ($GBS_{new} \geq GBS_{sofar}$) **then**
10:        break
11:    **else**
12:        $D = D'$
13:        $DIM = DIM'$
14:        $GBS_{sofar} = GBS_{new}$
15:    **end if**
16: **end loop**
17: Exit with D and GBS

Algorithm 2. The ABPM Variation

1: **for** $i = 1$ to $i = |DIM| - 1$ **do**
2:   **for** $j = i + 1$ to $j = |DIM|$ and $j \neq i$ **do**
3:     **for** $p = 1$ to $p = |K_i|$ **do**
4:       bs_{ij_p} = Banding score for index p in Dim_i calculated w.r.t. Dim_j using
            Equation 1
5:     **end for**
6:     $DIM' = Dim_i$ rearranged according to bs_{ij} values
7:     $D' = D$ rearranged according to DIM'_i
8:   **end for**
9: **end for**

---

pairing the banding score bs_{ij_p} for each index p in dimension Dim_i is calculated with respect to Dim_j (line 4). The calculated banding score values are then used to rearrange the indexes in dimension Dim_i (line 6) and consequently the data space D (line 8).

For the EBPM variation lines 6 to 7 in Algorithm 1 are replaced with the pseudo code given in Algorithm 3. As in the case of the ABPM variation, the EBPM algorithm iteratively loops over the data space calculating banding scores for each index p in each dimension Dim_i. For each dimension, the bs_{i_p} values are used to rearrange the indexes in the dimension (line 6 in Algorithm 1) which is then used to reconfigure D. Inspection of Algorithm 1 indicates that Eq. 2 is called repeatedly and on each occasion a set Max will be generated. It therefore makes sense to generate a collection of Max sets in advance and store these in a Maximum set Table (an M-Table). Each row in this table will represent one of the n dimensions to be rearranged. The length of each row will be equivalent to the largest number of dots associated with a single index in the dimension associated with the row.

Algorithm 3. The EBPM Variation

1: **for** $i = 1$ to $i = |DIM|$ **do**
2:   **for** $p = 1$ to $p = k_i$ **do**
3:     bs_{i_p} = Banding score for index p in Dim_i calculated using Eq. 2
4:   **end for**
5:   $DIM' = Dim_i$ rearranged according to bs_i values
6:   $D' = D$ rearranged according to DIM'_i
7: **end for**

5.1   Generation of the Set Max

In the foregoing, Eq. 2 requires a set Max, a set of maximum distances of potential dot locations to the origin. The size of the set max depends on the number of dot locations to be considered (the size of the set C in Eq. 2). The calculation of the longest possible distance from the origin to a dot within a n-D space is straight forward as the maximum coordinates are known. The second most longest distance is harder, especially where the ND space under consideration is not symetrical. Similarly with the third longest distance and so on. Other than for the maximum distance there will be a number of candidates locations that will give the nth most longest distance.

An algorithm for populating the set Max is thus given in Algorithm 4 (the Maximum Distance Calculation, or MDC, algorithm). The inputs to the algorithm are: (i) the number of maximum values to be returned (thus the size of the desired set Max) and (ii) the dimension sizes ($n-1$ because we exclude the current dimension for which banding scores are being calculated). The output is a list of maximum distances in descending order. On start up the location loc_1, which will feature the maximum distance, is identified and stored in the set Loc (line 5). The associated distance $dist_1$ is then calculated and stored in the set $Dist$ (line 7). The algorithm then continues, in an iterative manner, according to the $numValues$ input parameter. On each iteration the longest distance in $Dist$ is extracted and added to the list Max (lines 9 and 10). This distance is

---

Algorithm 4. Maximum Distance Calculation (MDC) Algorithm

1: **Input:** $numValues$ = The size of the desired set Max,
   $DimSizes = \{k_1, k_2, \ldots k_{n-1}\}$ (The sizes of the dimensions to be considered)
2: **Output:** Max = The desired set of maximum distances
3: $Max = \emptyset$
4: $loc_1 = \langle k_1, k_2, \ldots k_{n-1} \rangle$
5: $Loc = \{loc_1\}$ (Running list of locations)
6: $dist_1 = dist(loc_1)$
7: $Dist = \{dist_1\}$ (Running list of distances)
8: **for** $i = 1$ to $i = numValues$ **do**
9:     $j = getLongestDistIndex(Dist)$
10:     $Max = Max \cup Dist_j$
11:     $Dist = Dist - dist_j$ (Prune $dist_j$ from $Dist$)
12:     $Loc = Loc - loc_j$ (Prune loc_j from Loc)
13:     $NewLoc = \emptyset$
14:     **for** $q = 1$ to $q = |DimSizes|$ **do**
15:         $loc = loc_j$ with c_i replaced with $c_i - 1$
16:         **if** $loc \notin Loc$) **then**
17:             $NewLoc = NewLoc \cup loc$
18:         **end if**
19:     **end for**
20:     $NewDist = \emptyset$
21:     **for** $q = 1$ to $q = |NewLoc|$ **do**
22:         $dist = dist(newLoc_q \in NewLoc)$
23:         $NewDist = NewDist \cup dist$
24:     **end for**
25:     $Loc = Loc \cup NewLoc$
26:     $Dist = Dist \cup NewDist$
27: **end for**

then pruned from the set $Dist$ (line 11) and the associated location pruned from the set Loc (line 12). The algorithm then (lines 13 to 18) calculates a new set of locations, to be added to Loc, by iteratively subtracting 1 from each coordinate associated with Loc_j in turn ($loc_j = \langle c-1, c_2, \ldots, c_{n-1} \rangle$) and thus creating new location adjacent to loc_j. If not already in Loc each new location loc is appended to the set $NewLoc$. The algorithm then used the set $NewLoc$ to calculate a new set of distances $NewDist$ (lines 20 to 24). The sets $NewLoc$ and $NewDist$ are then appended to the existing sets Loc and $Dist$ and the process repeated until there are no more maximum distances to calculate. Note that the function $dist()$ in Algorithm 4 is the same as that used in Eq. 2.

6 Evaluation

This section presents the evaluation of the BPM algorithms considered in this paper. All the reported experiments were conducted using either: (i) 2-D data sets available within the UCI data mining repository [13] or (ii) 5-D data sets extracted from the Great Britain (GB) Cattle Tracking System (CTS). The latter was selected because it can be interpreted in the context of five dimensions. The objectives of the evaluation was firstly to compare the operation, in terms of the overall GBS and runtime (seconds), of the ABPM and EBPM algorithms, using both the Euclidean and Manhattan variations of the latter, in the context of n-D data sets (specifically 5-D data sets); and secondly to compare the operation, using an independent Average Band Width (ABW) measure as well as the overall GBS obtained, and runtime (in seconds), of the best BPM algorithm from (1) with the previously proposed BC and MBA algorithms.

6.1 Comparison of BPM Algorithms (ABPM and EBPM)

This section presents the results obtained with respect to the comparison of the BPM algorithms, founded on the banding score concept considered in this paper. The comparison was conducted by considering GBS and runtime. For the evaluation the operation of ABPM was compared with the operation of EBPM with either Euclidean or Manhattan distance calculation and with or without the use of M-Tables. The evaluation was conducted using the 16 5-D data sets extracted from the CTS database, each comprised of 24,000 records. The results with respect to the final GBS obtained are given in Table 1 and with respect to runtime (seconds) in Table 2. From Table 1 it can be observed that in all cases (as expected) the EBPM algorithm (both variations) produced better bandings than the ABPM algorithm. In addition Euclidean EBPM produced better bandings than Manhattan EBPM. The reason being that Euclidean distance measurement is more precise than Manhattan distance measurement and consequently Euclidean EBPM produced better bandings.

Table 1. Comparison of BPM algorithms in terms of GBS, best results in bold font.

County	Year	Num. Recs.	ABPM	EBPM Euclid.	Manhat.
Aberdeenshire	2003	24000	0.9066	**0.8502**	0.8615
	2004	24000	0.9096	**0.8381**	0.8584
	2005	24000	0.9265	**0.8364**	0.8551
	2006	24000	0.9119	**0.8408**	0.8846
Cornwall	2003	24000	0.8666	**0.8112**	0.8499
	2004	24000	0.8972	**0.8322**	0.8611
	2005	24000	0.8668	**0.8244**	0.8345
	2006	24000	0.8995	**0.8382**	0.8675
Lancashire	2003	24000	0.8984	**0.8112**	0.8562
	2004	24000	0.9121	**0.8443**	0.8617
	2005	24000	0.8839	**0.8276**	0.8377
	2006	24000	0.9001	**0.8452**	0.8457
Norfolk	2003	24000	0.9016	**0.8463**	0.8685
	2004	24000	0.8991	**0.8529**	0.8740
	2005	24000	0.9166	**0.8267**	0.8605
	2006	24000	0.9151	**0.8533**	0.8786
Average		24000	0.9007	**0.8362**	0.8597

6.2   Comparison with Previous Work (BC and MBA)

This section reports on the experiments conducted to compare the operation of the proposed BPM algorithm with the previously proposed the BC algorithm [24] and the two variations of MBA algorithm, MBA_{BFP} and MBA_{FP} [21]. The comparison was conducted in terms of efficiency (runtime measured in seconds) and effectiveness. Although the BPM algorithms considered in this paper seek to minimise a GBS, the BC and MBA algorithms operated in a different manner. The BC algorithm sought to maximise a Mean Row Moment (MRM) value, while the MBA algorithms sought to maximise an accuracy value. An independent measure was thus used for the comparison. More specifically the Average Band Width (ABW) measure was devised; the normalised average distance of dots from the diagonal measured according to the distances of the normals from the diagonal to each dot. ABW is calculated using Eq. 5, where: D is the set of dots (with each dot defined in terms of a set of cartesian coordinates) and $maxABW$ is the maximum possible ABW value given a particular data matrix size.

$$ABW = \frac{\sum_{i=1}^{i=|D|} distance\ d_i\ from\ leading\ diagonal}{|D| \times maxABW} \tag{5}$$

Table 2. Comparison of BPM algorithms in terms of runtime (seconds), best results in bold font.

County	ABPM	EBPM			
		With M-Table		Without M-Table	
		Euclid.	Manhat.	Euclid.	Manhat.
Aberdeenshire	**26.33**	36.85	30.27	75.41	63.61
	25.85	36.08	29.18	78.20	66.97
	22.35	34.42	28.42	75.31	60.93
	27.67	33.66	30.08	76.48	64.16
Cornwall	**25.20**	32.33	28.20	74.13	65.22
	26.75	38.86	30.84	73.13	63.38
	24.58	34.58	29.36	77.75	69.08
	23.49	36.38	30.31	72.20	67.45
Lancashire	**22.47**	39.21	28.01	79.59	63.03
	26.69	35.47	31.48	74.31	66.47
	25.45	32.25	28.94	69.16	59.34
	22.69	32.83	28.77	75.78	67.31
Norfolk	**26.91**	34.42	31.79	76.48	66.55
	22.91	35.70	27.72	73.88	62.52
	25.45	32.20	45.17	75.40	63.47
	24.99	33.07	29.78	69.22	61.55
Average	**24.99**	34.89	30.48	74.78	64.44

For the evaluation the operation of only the EBPM algorithm was compared with BC, MBA_{FP} and MBA_{BFP} (because in 2-D both variations of the EBPM algorithm and ABPM operate in the same manner, it is only in higher dimensions that there operation differs). Note also that because BC, MBA_{FP} and MBA_{BFP} were only designed to operate in 2-D the comparison was conducted using 2-D data sets taken from the UCI Machine learning repository processed so as to produce binary valued equivalents. In each case the attributes represented the x-dimension and the records the y-dimension. The results obtained are presented in Tables 3 and 4. Table 3 shows the effectiveness results obtained in terms of GBS and the independent ABW measure. From the table it can be observed that the EBPM algorithm produce better bandings, in terms of GBS and the independent ABW measure, than the other banding algorithms considered (best result highlighted in bold font). From Table 4 it can also be observed that the BC, MBA_{FP} and MBA_{BFP} algorithms all require considerably more processing time than the proposed EBPM algorithm. For completeness Fig. 2 shows the Lympography data set, before banding, and after banding using the EBPM algorithm.

Table 3. 2D banding evaluation results in terms of ABW and GBS, for the five banding mechanisms considered (best results in bold font)

Datasets	# Recs	# Cols	Before Banding ABW	GBS	EBPM ABW	GBS	MBA$_{FP}$ ABW	GBS	MBA$_{BFP}$ ABW	GBS	BC ABW	GBS
Adult	48842	97	0.4487	0.3662	**0.3318**	**0.1294**	0.4116	0.2869	0.3617	0.2539	0.3394	0.4005
ChessKRv	28056	58	0.4444	0.3473	**0.2208**	**0.1791**	0.3816	0.3699	0.3246	0.2832	0.3240	0.3997
LetRecognition	20000	106	0.4125	0.3325	**0.2885**	**0.1682**	0.3407	0.2751	0.3152	0.2632	0.3246	0.3965
PenDigits	10992	89	0.4276	0.3453	**0.2197**	**0.2064**	0.3318	0.2775	0.2872	0.2874	0.3276	0.4005
Waveform	5000	101	0.4372	0.3402	**0.2414**	**0.2091**	0.3774	0.2958	0.2951	0.3215	0.2833	0.3702
Mushroom	8124	90	0.4297	0.3473	**0.2638**	**0.1774**	0.3866	0.3284	0.3845	0.3018	0.3297	0.3448
Annealing	898	73	0.4433	0.4133	**0.3630**	**0.1218**	0.4389	0.3300	0.3779	0.3162	0.3826	0.2977
HorseColic	368	85	0.4009	0.3857	**0.3205**	**0.2367**	0.4001	0.3801	0.3881	0.3760	0.3353	0.2904
Heart	303	52	0.4346	0.4318	**0.3016**	**0.1502**	0.4142	0.3387	0.3338	0.2833	0.3423	0.2651
Wine	178	68	0.4430	0.4564	**0.2027**	**0.2785**	0.3645	0.3970	0.3061	0.4015	0.3384	0.2561
Hepatitis	155	56	0.4438	0.4619	**0.2957**	**0.2063**	0.3032	0.4240	0.2962	0.4279	0.3438	0.2629
Lympography	148	59	0.3356	0.4581	**0.2826**	**0.2487**	0.2887	0.4359	0.2804	0.4540	0.3324	0.2738
Average	10255	78	0.4251	0.3905	**0.2777**	**0.1927**	0.3699	0.3449	0.3292	0.3308	0.3336	0.3299

Table 4. Run-time (RT) Results (seconds) Using UCI data sets.

Data sets	# Rows	# Cols	runtime (secs) EBPM	BC	MBA$_{BFP}$	MBA$_{FP}$
Adult	48842	97	**76.74**	175.84	185.95	140.95
ChessKRvK	28056	58	**11.46**	23.27	27.90	27.81
LetRecognition	20000	106	**10.28**	26.38	24.54	21.31
PenDigits	10992	89	**02.81**	10.12	12.94	11.85
Waveform	5000	101	**0.88**	02.28	03.05	02.41
Mushroom	8124	90	**02.24**	08.47	09.07	08.14
Annealing	898	73	**0.05**	0.22	0.26	0.20
HorseColic	368	85	**0.02**	0.09	0.20	0.12
Heart	303	52	**0.02**	0.08	0.12	0.11
Wine	178	68	**0.01**	0.09	0.09	0.06
Hepatitis	155	56	**0.01**	0.08	0.06	0.06
Lympography	148	59	**0.01**	0.08	0.08	0.06
Average	10255	78	**08.71**	20.58	28.72	17.76

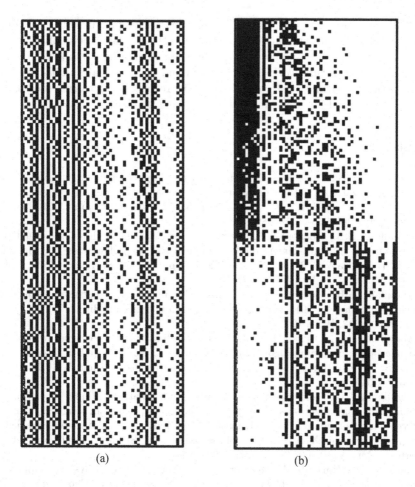

(a)   (b)

Fig. 2. Lympography data set, before (a) and after (b) banding using EBPM

7   Conclusion

A number of BPM algorithms have been presented. More specifically two alternative banding algorithms were considered, ABPM and EBPM. Four variation
of EBPM were considered, using Euclidean and Manhattan distance calculation,
and with and without the use of M-Tables. The presented evaluation established
the following main findings. The proposed BPM algorithms outperformed the
previously proposed BC, MBA$_{FP}$ and MBA$_{BFP}$ algorithms, with respect to 2-D
UCI data sets, in terms of: (i) efficiency, (ii) the GBSs obtained and (iii) an independent ABW measure. There is no difference in the operation of ABPM and
EBPM in 2-D. In higher dimensions (5-D data sets were considered) the ABPM
algorithm offered the advantage that it was more efficient than EBPM although
the quality of the bandings produced were not as good as those produced using

EBPM. In the context of EBPM the Euclidean variation produced the best quality bandings while the Manhattan variation was slightly faster. Use of M-Tables, to reduce the required amount of banding score calculation, was also found to be beneficial. For future work the authors intend to investigate multi-core variations of the algorithms that can be used with respect to platforms such as the Hadoop distributed file store and data processing platform.

References

1. Abdullahi, F.B., Coenen, F., Martin, R.: A novel approach for identifying banded patterns in zero-one data using column and row banding scores. In: Perner, P. (ed.) MLDM 2014. LNCS (LNAI), vol. 8556, pp. 58–72. Springer, Cham (2014). https://doi.org/10.1007/978-3-319-08979-9_5
2. Abdullahi, F.B., Coenen, F., Martin, R.: A scalable algorithm for banded pattern mining in multi-dimensional zero-one data. In: Bellatreche, L., Mohania, M.K. (eds.) DaWaK 2014. LNCS, vol. 8646, pp. 345–356. Springer, Cham (2014). https://doi.org/10.1007/978-3-319-10160-6_31
3. Abdullahi, F.B., Coenen, F., Martin, R.: Finding banded patterns in big data using sampling. In: 2015 IEEE International Conference on Big Data (Big Data), pp. 2233–2242. IEEE (2015)
4. Abdullahi, F.B., Coenen, F., Martin, R.: Finding banded patterns in data: the banded pattern mining algorithm. In: Madria, S., Hara, T. (eds.) DaWaK 2015. LNCS, vol. 9263, pp. 95–107. Springer, Cham (2015). https://doi.org/10.1007/978-3-319-22729-0_8
5. Abdullahi, F.B., Coenen, F., Martin, R.: Banded pattern mining algorithms in multi-dimensional zero-one data. In: Hameurlain, A., Küng, J., Wagner, R., Bellatreche, L., Mohania, M. (eds.) Transactions on Large-Scale Data- and Knowledge-Centered Systems XXVI. LNCS, vol. 9670, pp. 1–31. Springer, Heidelberg (2016). https://doi.org/10.1007/978-3-662-49784-5_1
6. Alizadeh, F., Karp, R.M., Newberg, L.A., Weisser, D.K.: Physical mapping of chromosomes: a combinatorial problem in molecular biology. Algorithmica 13(1–2), 52–76 (1995)
7. Atkins, J.E., Boman, E.G., Hendrickson, B.: Spectral algorithm for seriation and the consecutive ones problem. J. Comput. SIAM 28, 297–310 (1998)
8. Atkinson, K.E.: An Introduction to Numerical Analysis. John Wiley & Sons, New York (2008)
9. Aykanat, C., Pinar, A., Catalyurek, U.: Permuting sparse rectangular matrices into block-diagonal form. SIAM J. Sci. Comput. 25, 1860–1879 (2004)
10. Baeza-Yates, R., Ribeiro-Neto, B.: Modern Information Retrieval. Addison-Wesley, Reading (1999)
11. Bertin, J.: Graphics and Graphic Information Processing. Walter de Gruyter, New York (1981)
12. Bertin, J.: Graphics and graphic information processing. In: Readings in Information Visualization, pp. 62–65. Morgan Kaufmann Publishers Inc., (1999)
13. Blake, C.I., Merz, C.J.: UCI repository of machine learning databases (1998). www.ics.uci.edu/~mlearn/MLRepository.htm
14. Cheng, K.-Y.: Minimizing the bandwidth of sparse symmetric matrices. Computing 11(2), 103–110 (1973)

15. Cheng, K.-Y.: Note on minimizing the bandwidth of sparse, symmetric matrices. Computing **11**(1), 27–30 (1973)
16. Cheng, Y., Church, G.M.: Biclustering of expression data. In: ISMB, vol. 8, pp. 93–103 (2000)
17. Deutsch, S.B., Martin, J.J.: An ordering algorithm for analysis of data arrays. Oper. Res. **19**(6), 1350–1362 (1971)
18. Fortelius, M., Kai Puolamaki, M.F., Mannila, H.: Seriation in paleontological data using Markov chain monte carlo methods. PLoS Comput. Biol. **2**, 2 (2006)
19. Garriga, G.C., Junttila, E., Mannila, H.: Banded structures in binary matrices. In: Proceedings Knowledge Discovery in Data Mining (KDD 2008), pp. 292–300 (2008)
20. Garriga, G.C., Junttila, E., Mannila, H.: Banded structure in binary matrices. Knowl. Inf. Syst. **28**(1), 197–226 (2011)
21. Junttila, E.: Pattern in Permuted Binary Matrices. Ph.D thesis (2011)
22. Koebe, M., Knöchel, J.: On the block alignment problem. Elektronische Informationsverarbeitung und Kybernetik **26**(7), 377–387 (1990)
23. Liiv, I.: Seriation and matrix reordering methods: an historical overview. Stat. Anal. Data Min. **3**(2), 70–91 (2010)
24. Makinen, E., Siirtola, H.: The barycenter heuristic and the reorderable matrix. Informatica **29**, 357–363 (2005)
25. Mäkinen, E., Siirtola, H.: Reordering the reorderable matrix as an algorithmic problem. In: Anderson, M., Cheng, P., Haarslev, V. (eds.) Diagrams 2000. LNCS (LNAI), vol. 1889, pp. 453–468. Springer, Heidelberg (2000). https://doi.org/10.1007/3-540-44590-0_37
26. Mannila, H., Terzi, E.: Nestedness and segmented nestedness. In: Proceedings of the 13th ACM SIGKDD International Conference on Knowledge Discovery and Data Mining, pp. 480–489. ACM (2007)
27. Myllykangas, S., Himberg, J., Bohling, T., Nagy, B., Hollman, J., Knuutila, S.: DNA copy number amplification profiling of human neoplasms. Oncogene **25**, 7324–7332 (2006)
28. Sugiyama, K., Tagawa, S., Toda, M.: Methods for visual understanding of hierarchical system structures. IEEE Trans. Syst. Man Cybern. **11**, 109–125 (1981)

Automated Business Process Discovery and Analysis for the International Higher Education Industry

Juan Gonzalez-Dominguez[1]($\boxtimes$) and Peter Busch[2]

[1] AlphaSys Pty Ltd., Sydney, NSW 2000, Australia
`gonzalezd.juan@gmail.com`
[2] Department of Computing, Macquarie University,
Marsfield, NSW 2109, Australia
`peter.busch@mq.edu.au`

Abstract. The international education sector in Australia expects a rise in demand for higher education programs in the future. This increase drives universities to monitor and continuously improve their business processes. Automated Business Process Discovery or Process Mining helps organisations create a business process graphical representation or process model, leveraging data from their information systems. The lack of a business process model and inefficiency in harnessing process data deters fact-based process improvement decision-making, impacting a company's operations. Within the international education context, Process Mining can help discover, analyze, manage and improve processes efficiently and deal with increasing market demand of international student enrolments. Here we have used the Process Mining Methodology Framework for Discovery Analysis focusing on Control-flow and Case Data perspectives. We highlight challenges encountered in process data extraction and preparation as well as performance analysis to identify potential bottlenecks.

Keywords: Automated business process discovery · Process mining Business process · Process modeling

1 Introduction

Business Process Management (BPM) uses different techniques to achieve organisational goals through well established, monitored and continuously improving processes [1]. One of these techniques is *Automated Business Process Discovery* or *Process Mining*. Process mining discovers business process representations (or process models) using data mining techniques [2]. Furthermore, process mining can help identify process deviances and support business process improvement initiatives [3]. The data for process mining is recorded in Process-Aware Information Systems (PAIS) owned by organisations to perform daily activities [4]. The data is extracted in the form of Event Logs [4]. Within the process mining body of knowledge, the Process Mining Manifesto covers guiding principles and challenges that need to be addressed [3]. Process mining could also be implemented to achieve different goals, for example,

© Springer Nature Switzerland AG 2018
K. Yoshida and M. Lee (Eds.): PKAW 2018, LNAI 11016, pp. 170–183, 2018.
https://doi.org/10.1007/978-3-319-97289-3_13

basic process discovery, case data analysis or organisational analysis. In-depth analysis can be performed to achieve conformance and performance analysis [5]. Process mining approaches can be applied to different industries, for instance, government and financial institutions [5, 6]. The international education industry in Australia, more specifically the higher education sector could harness process mining approaches to cope with an increase in future enrolment demand predicted by *Deloitte Access Economics* [7] and continuously improve processes. Here we present knowledge obtained by applying process mining techniques to discover a sample *International Students Admission Process*.

2 Background

Business processes are a series of interdependent activities carried out at different levels. Processes can run internally in an organisation, either inside a department or among various departments, or externally through different organisations [8]. Several authors have approached the study of business processes from different perspectives. Papazoglou and Ribbers [9] describe business processes as part of the technical and management foundations for e-business solutions. Hammer [10] notes processes and process management as a whole discipline supports transformation strategies and organisational operations. Hung [11] notes process alignment and people involvement are variables impacting the successful implementation of business process management as a competitive advantage. Business process initiatives allow organisations to develop or implement methodologies to maximise resources by business process continuous improvement [12].

The use of data mining methodologies for automated business process discovery (ABPD) has increased in research [2]. The main goal is to leverage existing data recorded at the execution of activities within process-aware information systems (PAIS) to create a business process graphical representation or business process model [2]. Automated business process discovery is also known as Process Mining. In this area, Professor Will van der Aalst is recognised as one of the major contributors to the process mining body of knowledge. His publications cover a range of process mining areas, e.g. workflow mining [13], obtaining data [4], differing techniques to deal with event logs [14] and the Process Mining Manifesto [3].

Mans et al. [15] and Jutten [16] explain there is considerable academic literature about process mining, but research about adoption is infrequently found, for organisations do not foresee business benefits from this technique. There are some documented case studies, e.g. a provincial office of the Dutch National Public Works Department provided different standpoints for the use of process mining for process discovery and performance [6], and business alignment or conformance checking [17]. In De Weerdt et al. [5] a detailed case study in a financial institution shows the capabilities of process mining in real-life environments, but the authors conclude the need to focus on practical applicability to improve process mining techniques. But what is process mining?

3 Process Mining Overview

Since the publication of the value chain by Porter [18] organisations have shifted from function-centric to process-centric operations. Business processes are the collection of interdependent activities performed in an established order to reach organisational goals [8]. Several measurements can be attached to business processes, for instance, time, cost, performer and quality. Furthermore, business processes' scope can range from processes running within an individual business unit to processes running across different organisations [9], but they typically require some form of management.

3.1 Business Process Management

Business Process Management (BPM) is considered a mature discipline that "supports business processes using methods, techniques, and software to design, enact, control and analyse operational processes involving humans, organisations, applications, documents and other sources of information" [1]. BPM uses techniques such as business process modelling to help create representations of business processes (also known as business process models), for better visualization and to avoid misunderstandings [19]. BPM can use tools and techniques from different approaches. Methodologies based on creativity like Business Process Re-engineering (BPR), internal or external benchmarking or statistical analysis such as Six Sigma [20] are also relevant. For example van der Aalst et al. [21] highlight how Six Sigma methodologies for improving a business process by "statistically quantifying process performance" [21] could be not accurate as the data would be manually collected, as well as prove expensive and time-consuming. *Automated business process discovery* better known as Process Mining can overcome this issue.

3.2 Process Mining

Process mining is a technique depicting business process models through patterns in big data sets comprising data recorded at process run time in information systems; it originates from the adjustment and growth of data mining techniques applied to the field of business processes [2]. "Discover, monitor and improve real processes (i.e., not assumed processes) by extracting knowledge from event logs readily available in today's (information) systems" [3] is process mining's goal. Process mining makes a significant distinction between real processes and assumed processes. The former relates to the creation of process models based on tasks recorded in information systems – i.e. actions that really happened. The latter refers to the creation of process models based on human observations, expectations and assumptions [22, 23].

In 2009 an IEEE Process Mining Task Force was created, and in 2011 it released the final version of the Process Mining Manifesto [24]. The manifesto describes six guiding principles (GP) that explain the importance of data (event logs), extraction, how to treat the depicted process model as an abstraction of reality and process mining as a continuous activity [3]. These principles are: GP1 - *Event data should be treated as First-Class Citizens* [24]. GP2 - *Log Extraction should be driven by questions* [24]. GP3 - *Concurrency, choice and other basic control-flow constructs should be supported* [24]. GP4 - *Event should be related to model elements* [24]. GP5 - *Models should be treated as a purposeful abstraction of reality* [24]. GP6 - *Process Mining should be a continuous process* [24].

3.3   Event Logs

Process data plays the most important role in process mining. This data is often extracted and converted into event logs, representing the starting point for process mining [24]. For process mining implementations as discussed in GP4, event logs shall only contain event data related to the process under analysis [4]. An event log can be deconstructed to the following elements: *Cases*, which represent a process instance, therefore the event log would contain several cases. *Events*, every case is formed by events, these could be understood as a task in the process and every event is part of one and only one case. Event *attributes* - any extra information related to the process. Common attributes are activity, timestamp, cost and resources [4]. Figure 1 represents an example event log:

Case id	Event id	Properties			
		Timestamp	Activity	Resource	Cost ...
1	35654423	30-12-2010:11.02	Register request	Pete	50 ...
	35654424	31-12-2010:10.06	Examine thoroughly	Sue	400 ...
	35654425	05-01-2011:15.12	Check ticket	Mike	100 ...
	35654426	06-01-2011:11.18	Decide	Sara	200 ...
	35654427	07-01-2011:14.24	Reject request	Pete	200 ...

Fig. 1. A sample event log (Source: van der Aalst, 2011: 99)

Challenges for event logs are covered by the Process Mining Manifesto. Ly et al. [25] noted four of them: incorrect/incomplete log data, data contribution through parallel branches, infrequent traces (planned exceptions) and ad-hoc contributed data (unplanned exceptions. Just as an aside, GP1 establishes the level of maturity for an event log. The maturity level considers four criteria for measurement Trustworthiness, Completeness, Well-defined Semantics and Security.

3.4   Types of Process Mining

Process Mining may be divided into three types. *Discovery* - here event logs are used to create a graphical representation of the business process. This graphical representation is also known as *process model*. The main output for *discovery* is a business process model. Discovery is the main application of process mining in organisations. *Confor-mance* - here an event log is compared to an existing business process model. The aim is to corroborate if the process recorded in the event log (real activities) runs as expected by the preliminary model. The primary output in *conformance* is process model diagnosis. *Enhancement* - the goal is to improve the current process model by using information in the event logs. In contrast with *conformance*, instead of comparing with established measurements, in *enhancement* the idea is to change and improve the process model. The primary output in *enhancement* is a new business process model [3].

3.5   Process Mining Methodology Framework

The process mining follows methodologies according to the aim of the implementation, for instance, Bozkaya et al. [26] propose five phases for process diagnostic with process mining, log preparation, log inspection, control flow analysis, performance analysis and role analysis. Albeit this methodology has been applied to government and financial organisations [26, 27], it does not cover all the process mining perspectives.

De Weerdt et al. [5] developed the Process Mining Methodology Framework (PMMF). This methodology can be considered more flexible, including an analysis phase where different perspectives can be considered depending on the process mining aim. PMMF comprises five phases. The first phase is *preparation*. In this phase data is extracted from information systems and prepared as event logs. Next *extraction* takes into consideration the scope and timeframe that best fits the process mining implementation. An excess of information can result in extraction of activities not belonging to process scope. A timeframe not covering possible seasonal behavior, will not be represented in the process discovery [5]. The next phase - *exploration* is an iterative activity where scope and timeframe are analysed and re-configured based on multiple process visualisations with different algorithms. This phase holds a relationship with the activity of data extraction as exploratory outcomes can show the need for more data or different time frames [5]. In the *perspectivization* phase, the perspective the analysis will pursue is determined and if it is necessary to construct different event logs for each perspective [5]. The next phase is *analysis* and could be divided in two - *Discovery analysis* and *in-depth analysis*. Basic *discovery analysis* is where various dimensions recorded in event logs are analysed. *Control-flow analysis* refers to the sequence of activities. *Organisational analysis* relates to the users' relations and social networks created by the process run time. *Case data analysis* refers to other attributes included, for example cost and time [5]. With *in-depth analysis* more detailed examination takes place. *Conformance analysis* would help identify activity deviations from the expected process model based on the mined process model. *Performance analysis* supports organisations to find insights related to the process, for instance process execution times and waiting periods [5]. The final phase is *results* - in this phase all findings during the analysis phase are considered for decision making and continuous business process improvement [5]. Let us now examine a working scenario to place the above in to context.

4   Process Mining Knowledge Generation

With growing demand from international students, a university should have the continuous capability to monitor and improve its *International Students Admissions Process*. The importance of this process relies on the ability of a university to (a) ensure candidates' qualifications and English proficiency are appropriate, (b) comply with the Australian Education Legislation of providing written agreement [28], (c) facilitate the transition between the application stage and the student enrolment stages and

(d) articulate market development strategies for recruitment. To manage the process, a university may use an application system recording all daily transactions regarding admission activities and helping to keep one source of information. In general, the process is triggered when an application for study is received. The application can be made online through the application system or in hard copy (paper application). Data entry clerks can review the application and create a new student record in the student management system. If the application is paper-based, student details are entered in two systems. The application can then be turned over for assessment to the admissions assistant or admissions officers.

During assessment, admission officers confirm if the documentation is complete, if it complies with university policies and if the student meets the entry requirements for the desired program. Furthermore, faculty permission and recognition of prior learning advice are sought if necessary. After assessment, admissions officers either reject the application, ask for more information, or issue an offer letter with all the information about the program, tuition fees, commencement dates, insurance and other information requested by Australian legislation [28]. The offer letter can be a full offer letter, full offer with conditions or conditional. The next task is acceptance. A significant time gap can be expected between issuing the offer letter and acceptance. The acceptance is processed by the admissions officer with assistance in coordination from the finance department. When the finance department confirms the student's payment, a confirmation of enrolment (CoE) is issued and sent to the student with a welcome message and instructions for program enrollment. At this point, the *International Students Admissions Process* can be considered finished. A theoretical example of the *International Students Admission Process* model can be *discovered* by extracting an event log from an application system. Remember what we seek here is process discovery, following the PMMF for control flow and case analysis.

4.1 Event Log Extraction

A system records all the activities related to the admissions process including application status, any action performed in the system, the name of the person executing activities, documents, a student's demographic information and more. The system's outputs are reports in comma separated value format extracted by the reporting team. Every day an internal student report with all students in the system is extracted, this report only represents a static picture of a student's status at the moment of reporting. The system is a stand-alone solution with no implementation of an Application Programming Interface (API) for extraction of student information. Due to the aforementioned reason, extraction must be made manually from each student activity log, taking into consideration the scope and timeframe.

As a working example, we focus only on complete process instances [5]. Based on a generated daily report, a sample of applications with "Acceptance", "Withdrawn" and "Not Qualified" was created. These statuses are considered termination points for the admissions process. A sample calculation was made with 95% confidence, 5% of margin error, 50% distribution and a finite population correction applied [29]. Randomly generated samples by status were: 198 over 825 "not qualified", 243 over 1306 "withdrawn" and 294 over 2486 "acceptance" for a total of 735 cases.

4.2   Event Log Preparation

After extraction of data, an event log has to be explored to understand the data that is on it and verify if it will help to achieve the project goals [30]. That is to say, either eliminate data not useful for analysis or include data from other sources to complement the analysis [30]. Furthermore, any additional transformation such as de-identification of personal data was applied in this phase.

The event log extracted was enhanced with extra attributes that will help with the *Discovery Analysis for Case Perspective*. Data such as timestamp, person starting the event (*originator*), country of citizenship, application made by an agent or directly, and type of program, are some of the attributes added to the event log. An event log is stored in .csv format. To be used, it should be transformed to a MXML or XES (eXtensible Event Stream) format. XES was chosen as being the most current standard, is also more flexible to handle extra attributes in the event log and a plug-in is already available in the ProM [31].

4.3   Automated Business Process Discovery

To create a business process model based on the data provided by a sample event log, process mining makes use of algorithms, mainly data mining clustering algorithms [32]. For this experiment a heuristic miner was used. The algorithm analyses an event' "direct dependency, concurrency and not-directly-connectedness" [33] frequency, can handle short loops and is recommended for semi-structured processes [33]. The heuristic algorithm supports control-flow constructors like sequence, parallelism, choices, loops, invisible tasks and in some degree non-free-choice [34]. It also provides a middle point between underfitting and overfitting process models allowing fitness, simplicity, precision and generalisation criteria mentioned in the *Process Mining Challenge VI* [34]. In contrast, an α-algorithm requires structured processes, complete logs with no noise and no loops [33]. Albeit there have been some improvements in this algorithm such as α+ and α++, they still lack flexibility [33]. According to De Weerdt, De Backer, Vanthienen and Baesens [33] a Fuzzy Algorithm is preferred for event logs with a high level of noise and unstructured processes.

A generated event log for *discovery* contained 726 cases, 7,401 events, and 26 classes (status). By using a heuristic algorithm, there was potential to mine the *International Student Admission Process*. Figure 2 depicts the discovered process model. The discovered process shows certain tasks with higher frequency such as new application (starting task), submitted, pending assessment, assessing and application edited. Notable are also a start activity and multiple end activities, as well as several loops found in the event log.

The mined process shows the most common flow - new application - submitted - pending assessment - assessing – qualified - acceptance pre-processing - acceptance, achieving the fitness criteria. It also depicts possible paths the process can follow achieving precision. Note that simplicity and generalisation can be improved. One option can be reducing the process scope or by creating concept hierarchies for assessing or agent assignment activities.

The process illustrated in Fig. 2 comprises 26 different possible tasks performed during execution. The top five activities with more occurrences are assessing,

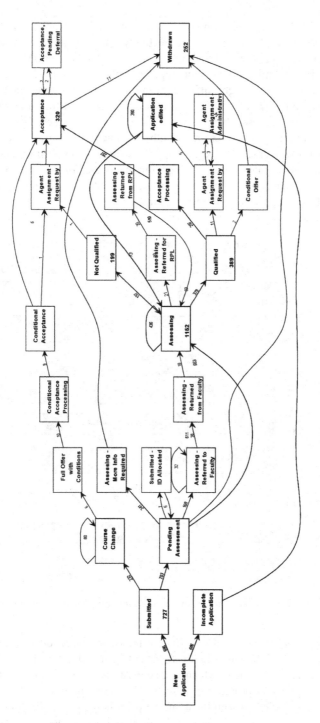

Fig. 2. A sample data-mined *International Student Admission Process*

application edited, submitted, pending assessment, and incomplete application with 1152, 909, 727, 711, 689 occurrences respectively. The mined process considers "new application" as the main starting activity (94.5% of the instances). However, some cases start with "submitted" and "incomplete application" both 5.5% of the cases. Regarding the end events, eight different events were found - acceptance, withdrawn, not qualified making 97.38% of the total cases and identified from the process scope. Application edited, course change, agent assignment – request by student, agent assignment – administrative error are the other end activities shown in the discovered process model. On average a process case can change status 9 times and include 10 activities from beginning to end. The maximum number of activities is 29 and the minimum is 2. The control flow analysis discovered 476 different process paths. Figure 3 included the ten most common paths:

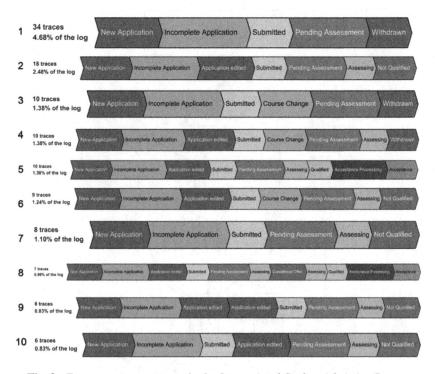

Fig. 3. Ten most common traces in the *International Student Admission Process*

These 10 paths represent only 16.26% of the entire paths found in the event log. Many of them end either with withdrawn or not qualified. Paths 1 and 3 were withdrawn without any assessment, which can suggest these applications do not fulfil the department's application processing policies. The mined process in Fig. 2 shows two activities that do not accord with the assumed process flow. "Application edited" is an activity that can happen in any part of the process but it happens mainly before the application is submitted. In the process model, this activity leads to "assessing" -

skipping two activities "submitted" and "pending assessment". The other activity is "not qualified" - considered as an ending activity. However one case in the event log continues with "agent assignment request by student" which allows the algorithm to link the activities - this case can be an exception within the process.

The goal of the *performance analysis* is to identify activities that impact the process flow negatively. Performance analysis within ProM was achieved by mining a Petri Net. To include all the possible task and the time calculations, the process model was overfitted for the 726 cases. A performance and conformance checking plugin was then applied. Table 1 summarises the main events and processing times per sequence pattern

Table 1. Summary events transitions processing times in days

Event from	Event to	Avg.	St. Dev.
New application	Submitted	3.2	18.09
Submitted	Pending assessment	0.9	1.01
Pending assessment	Assessing	4.2	7.16
Assessing	Assessing - more info required	1.7	7.42
Assessing	Assessing - referred for RPL	9.5	12.37
Assessing	Assessing - referred to faculty	6.3	12.92
Assessing - referred to faculty	Assessing - returned from faculty	21.5	14.94
Assessing - referred for RPL	Assessing - returned from RPL Assessment	6.8	10.64
Assessing	Qualified	11.3	18.58
Assessing - more info required	Qualified	23.0	20.81
Assessing - returned from faculty	Qualified	8.3	9.32
Assessing - returned from RPL assessment	Qualified	9.7	10.25
Assessing	Conditional offer	8.6	14.18
Assessing - more info required	Conditional offer	18.5	16.67
Assessing - returned from faculty	Conditional offer	2.3	1.78
Assessing - returned from RPL assessment	Conditional offer	6.2	4.36
Assessing	Not qualified	17.2	24.32
Assessing - more info required	Not qualified	32.8	24.72
Assessing - returned from faculty	Not qualified	11.3	22.88
Submitted	Withdrawn	20.8	27.10
Conditional offer	Withdrawn	32.9	26.75
Full offer with conditions	Withdrawn	82.0	27.80
Qualified	Withdrawn	33.2	27.88
Assessing	Withdrawn	26.2	29.28
Pending assessment	Withdrawn	20.2	26.70
Conditional offer	Qualified	21.2	18.36
Qualified	Acceptance processing	19.0	13.92
Acceptance processing	Acceptance	1.0	3.28
Full offer with conditions	Conditional acceptance processing	20.0	18.61
Conditional acceptance processing	Conditional acceptance	1.4	4.08

in days. The table shows average times and standard deviation. The latter is only indicative of the occurrences distribution (the range where most of the observations fall). It can be assumed the data has a positively skewed distribution, as the goal is to maintain the response times as low as possible. ProM does not provide median and quartile data to make a more accurate evaluation [35].

Overall results for the complete sample data set mentioned an average of 1.06 months to complete a case with a standard deviation of 1.02 months. The maximum time of processing is 9.61 months and the observation period was 11.69 months. Applications can be reviewed and entered into the student management system in less than one day and then hold a pending assessment status for 4.2 days. If advice is requested from faculties or recognition of prior learning, this could take 21.5 and 6.8 days respectively. Issuing an offer or rejection letters when documentation is complete and no further faculty or recognition of prior learning is sought, could be done in an average of 12.4 days (including qualified, conditional offers and not qualified status).

It was determined four activities increased processing time - *pending assessment, refer to faculty, refer for RPL and assessment* and *more information required. Pending assessment* relies completely on the admissions department and has a waiting time of 4.2 days. The latter three activities are delegated to external actors, meaning the admissions department cannot control the waiting time. Waiting time for faculty response is determined to 21.5 days. When the application required more information from the student, a total of 24.8 days to reply could elapse. The RPL team replied in 6.8 days on average.

5   Discussion

Process mining applications have been documented in academic research in different domains, such as government and financial services [5, 6]. Here process mining approaches were used in a university domain to obtain a theoretical process model for the *International Students Admission Process*. Process mining leverages data recorded in an application system to yield a business process model and information about the process performance. Some challenges and limitations were observed. The mined process model failed to achieve the *simplicity* and *generalisation* quality criteria as application processes can be very flexible and activities within the flow can be repeated at any given time, creating short loops. Due to this reason, the heuristic algorithm mapped all possible paths determined by the dependencies and occurrences, resulting in a complex process model with multiple flows – i.e. a process not simple to explain and general enough to cover all possible process flows.

The discovery analysis also detected some process instances that did not conform to what was assumed. *Starting* and *ending* events deviated from expected events. These process instances can be considered exceptions, but further analysis can be pursued to determine the original cause and authorizations. During data exploration, it was realised manual activities or activities not performed within the application system would not be included in the analysis. For instance, lead time between paper applications received and data entry into the system and issue escalation for special approvals, is occasionally not documented in the system. Some specific flows ended in "withdrawn" applications without any assessment. Perhaps causes for these cases could be reviewed and evaluated if the system can help reduce these occurrences or put in place business rules, business process or policies to reduce the workload generated from these instances.

Performance analysis identified potential activities that could be reviewed further. Most of these involve collaboration with external entities. More detailed assessment of these activities is recommended to address the reduction in processing time in a collaborative fashion. Regarding *assessing – more information required*, this is an activity relating directly to the student submitting incomplete applications. Understanding the root causes can help to elaborate solutions to solve potential issues, for instance information accessibility. The performance analysis also provided a general overview of processing times. Although processing times appeared within normal processing times, goals and strategies need to be put in place to consider the increasing demand leading to bottleneck activities described previously. Further work is required to achieve additional meaningful insights about the process. Mining specific flows separately, such as flows ending in acceptance, not qualified or withdrawn could offer more detailed information about performance and an applicants' characteristics to implement business process improvement or data mining initiatives. Finally, ProM is a powerful tool for process mining; it contains plugins providing functionality for process discovery, analysis and transformation [36]. While the software includes comprehensible user interfaces, it requires a certain type of process, data types, document formats, data mining and process mining understanding; hence non-expert users could find difficulties in using this tool. ProM could improve workspace management, particularly how results can be stored to avoid repetition of tasks for result gathering.

6 Conclusion

Automated Business Process Discovery or Process Mining is a technique for business process discovery, analysis and improvement [24]. This project implemented process mining approaches for control-flow and performance analysis [5] via a theoretical enrollment case. A business process model was discovered achieving fitness and precision quality criteria [3]. Non-standard activities for process start and end were also uncovered. Performance analysis identified process bottlenecks in four tasks - *refer to faculty, refer for RPL, assessing – more information required* and *pending assessment*. Further analysis will almost certainly produce better process insights. We recommend detailed process mining discovery and performance be applied to detailed flows. Other case perspective variables can be included for analysis as well.

References

1. van der Aalst, W.M.P., ter Hofstede, A.H.M., Weske, M.: Business process management: a survey. In: van der Aalst, W.M.P., Weske, M. (eds.) BPM 2003. LNCS, vol. 2678, pp. 1–12. Springer, Heidelberg (2003). https://doi.org/10.1007/3-540-44895-0_1
2. Tiwari, A., Turner, C.J., Majeed, B.: A review of business process mining: state-of-the-art and future trends. Bus. Process Manag. J. **14**, 5–22 (2008)

3. van der Aalst, W., et al.: Process mining manifesto. In: IEEE Task Force on Process Mining, pp. 1–19 (2012)
4. van der Aalst, W.M.P.: Getting the data. In: van der Aalst, W.M.P. (ed.) Process Mining: Discovery, Conformance and Enhancement of Business Processes, pp. 95–123. Springer, Heidelberg (2011). https://doi.org/10.1007/978-3-642-19345-3_4
5. De Weerdt, J., Schupp, A., Vanderloock, A., Baesens, B.: Process Mining for the multi-faceted analysis of business processes—a case study in a financial services organization. Comput. Ind. **64**, 57–67 (2013)
6. van der Aalst, W.M.P., Reijers, H.A., Weijters, A.J.M.M., van Dongen, B.F., Alves de Medeiros, A.K., Song, M., Verbeek, H.M.W.: Business process mining: an industrial application. Inf. Syst. **32**, 713–732 (2007)
7. Deloitte Access Economics, Growth and Opportunity in Australian International Education, pp. 1–95. Austrade, Melbourne (2015)
8. Davenport, T.H., Short, J.E.: The new industrial engineering: information technology and business process redesign. Sloan Manag. Rev. **31**, 11 (1990)
9. Papazoglou, M.P., Ribbers, P.: e-Business: Organizational and Technical Foundations. Wiley, Chichester (2006)
10. Hammer, M.: What is business process management? In: Vom Brocke, J., Rosemann, M. (eds.) Handbook on Business Process Management, pp. 3–16. Springer, Heidelberg (2010). https://doi.org/10.1007/978-3-642-00416-2_1
11. Hung, R.Y.-Y.: Business process management as competitive advantage: a review and empirical study. Total Qual. Manag. Bus. Excell. **17**, 21–40 (2006)
12. Harrington, H.J.: Business Process Improvement: The Breakthrough Strategy for Total Quality, Productivity, and Competitiveness. McGraw-Hill, New York (1991)
13. van der Aalst, W.M.P., Weijters, T., Maruster, L.: Workflow mining: discovering process models from event logs. IEEE Trans. Knowl. Data Eng. **16**, 1128–1142 (2004)
14. van der Aalst, W.M.P.: Extracting event data from databases to unleash process mining. In: vom Brocke, J., Schmiedel, T. (eds.) BPM - Driving Innovation in a Digital World, pp. 105–128. Springer, Cham (2015). https://doi.org/10.1007/978-3-319-14430-6_8
15. Mans, R., Reijers, H., Berends, H., Bandara, W., Rogier, P.: Business process mining success. In: Proceedings of the 21st European Conference on Information Systems, AIS Electronic Library (AISeL), Utrecht University, The Netherlands (2013)
16. Jutten, M.G.: The Fit Between Business Processes and Process Mining Related Activities: A Process Mining Success Model (2015)
17. van der Aalst, W.: Business alignment: using process mining as a tool for Delta analysis and conformance testing. Requir. Eng. **10**, 198–211 (2005)
18. Porter, M.E.: The Value Chain and Competitive Advantage (Extract). Free Press, New York (1985)
19. Burattin, A.: Introduction to business processes, BPM, and BPM systems. In: Burattin, A. (ed.) Process Mining Techniques in Business Environments: Theoretical Aspects, Algorithms, Techniques and Open Challenges in Process Mining. LNBIP, vol. 207, pp. 11–21. Springer, Cham (2015). https://doi.org/10.1007/978-3-319-17482-2_2
20. Siha, S.M., Saad, G.H.: Business process improvement: empirical assessment and extensions. Bus. Process Manag. J. **14**, 778–802 (2008)
21. van der Aalst, W.M.P., La Rosa, M., Santoro, F.M.: Business process management. Bus. Inf. Syst. Eng. **58**, 1–6 (2016)

22. van der Aalst, W.M.P.: Trends in business process analysis: from verification to process mining. In: Cardoso, J., Cordeiro, J., Filipe, J. (eds.) Proceedings of the 9th International Conference on Enterprise Information Systems (ICEIS 2007), pp. 12–22. Institute for Systems and Technologies of Information, Control and Communication, INSTICC, Medeira, Portugal (2007)

23. Medeiros, A.K.A.: Process mining: extending the α-algorithm to mine short loops. In: Beta, Research School for Operations Management and Logistics (2004)

24. van der Aalst, W.M.P.: Process mining: overview and opportunities. ACM Trans. Manag. Inf. Syst. (TMIS) **3**, 1–17 (2012)

25. Ly, L.T., Indiono, C., Mangler, J., Rinderle-Ma, S.: Data transformation and semantic log purging for process mining. In: Ralyté, J., Franch, X., Brinkkemper, S., Wrycza, S. (eds.) CAiSE 2012. LNCS, vol. 7328, pp. 238–253. Springer, Heidelberg (2012). https://doi.org/10.1007/978-3-642-31095-9_16

26. Bozkaya, M., Gabriels, J., van der Werf, J.M.: Process diagnostics: a method based on process mining. In: 2009 International Conference on Information, Process, and Knowledge Management, pp. 22–27 (2009)

27. Jans, M., van der Werf, J.M., Lybaert, N., Vanhoof, K.: A business process mining application for internal transaction fraud mitigation. Expert Syst. Appl. **38**, 13351–13359 (2011)

28. Department of Education Employment and Workplace Relations, National Code of Practice for Registration Authorities and Providers of Education and Training to Overseas Students 2007. In: Australian Government (ed.) ACT, pp. 1–34 (2007)

29. Agresti, A., Franklin, C.A. (eds.): Statistics: The Art and Science of Learning from Data, 2nd edn. Pearson Prentice Hall, Upper Saddle River (2009)

30. Medeiros, A.K.A., Weijters, A.J.M.M.: ProM Framework Tutorial, p. 41. Technische Universiteit Eindhoven, Eindhoven, The Netherlands (2009)

31. Verbeek, H.M.W., Buijs, J.C.A.M., van Dongen, B.F., van der Aalst, Wil M.P.: XES, XESame, and ProM 6. In: Soffer, P., Proper, E. (eds.) CAiSE Forum 2010. LNBIP, vol. 72, pp. 60–75. Springer, Heidelberg (2011). https://doi.org/10.1007/978-3-642-17722-4_5

32. Burattin, A.: data mining for information system data. In: Burattin, A. (ed.) Process Mining Techniques in Business Environments: Theoretical Aspects, Algorithms, Techniques and Open Challenges in Process Mining. LNBIP, vol. 207, pp. 27–32. Springer, Cham (2015). https://doi.org/10.1007/978-3-319-17482-2_4

33. De Weerdt, J., De Backer, M., Vanthienen, J., Baesens, B.: A multi-dimensional quality assessment of state-of-the-art process discovery algorithms using real-life event logs. Inf. Syst. **37**, 654–676 (2012)

34. van Dongen, B.F., Alves de Medeiros, A.K., Wen, L.: Process mining: overview and outlook of petri net discovery algorithms. In: Jensen, K., van der Aalst, W.M.P. (eds.) Transactions on Petri Nets and Other Models of Concurrency II: Special Issue on Concurrency in Process-Aware Information Systems. LNCS, vol. 5460, pp. 225–242. Springer, Heidelberg (2009). https://doi.org/10.1007/978-3-642-00899-3_13

35. Han, J., Kamber, M. (eds.): Data Mining: Concepts and Techniques, 2nd edn. Elsevier/Morgan Kaufmann, Amsterdam, Boston/San Francisco (2006)

36. van der Aalst, W.M., Van Dongen, B.F., Günther, C.W., Mans, R.S., de Medeiros, A.A., Rozinat, A., Song, M., Verbeek, H.M.W., Weijters, A.J.M.M.: Process mining with ProM. In: Proceedings of the 19th Belgium-Netherlands Conference on Artificial Intelligence, Belgium-Netherlands (2007)

An Analysis of Interaction Between Users and Open Government Data Portals in Data Acquisition Process

Di Wang[1,2(✉)] [iD], Deborah Richards[2], and Chuanfu Chen[1]

[1] Wuhan University, Wuhan 430072, Hubei, China
perditawd@gmail.com, cfchen@whu.edu.cn
[2] Macquarie University, Sydney, NSW 2109, Australia
deborah.richards@mq.edu.au

Abstract. The rate of development of open government data (OGD) portals has been fast in recent years due to potential benefits of the utilization of OGD. However, scholars have emphasized lack of use as a key problem in the realization of these benefits. Although studies have been carried out to understand decisive factors in OGD utilization from the aspects of either portals or users, they failed to consider the interaction between the two. Therefore, our study carried out an analysis of the interaction between users and OGD portals during users' data acquisition process from three aspects: data acquisition methods, data quality requirements, and helping functions. We carried out a survey in a Chinese population to collect data for analysis. Results show users' high acceptance of keyword search as their method for data acquisition through OGD portals but browsing showed higher usage frequency and was a more stable data acquisition behavior. Females show better acceptance of regular recommendations (e.g. RSS) based on their visiting histories than males. Users' age, education background and occupation affect their demands of different data quality attributes. Our analysis also shows positive relationship between users' data acquisition habits with their demands of data quality, users' need of help with their feelings of difficulties in using the portal, and users' need of help with their demands of data quality. We suggest promoting OGD utilization by offering better helping functions and improving data qualities in future development of OGD portals.

Keywords: Open government data · Data portal · Data acquisition

1 Introduction

In recent years, many open government data (OGD) portals have been built to publish data owned by the government to the public [1]. Due to the functions and abilities of these portals [2], they have been treated as flagship initiatives of OGD [3]. However, releasing large amounts of OGD to the public through portals is not a guarantee for the achievement of OGD initiatives in their targeted aims [1, 4] to promote citizen's engagement [5] and governments' transparency and accountability [1].

© Springer Nature Switzerland AG 2018
K. Yoshida and M. Lee (Eds.): PKAW 2018, LNAI 11016, pp. 184–200, 2018.
https://doi.org/10.1007/978-3-319-97289-3_14

On the contrary to the constant development of OGD portals, scholars have emphasized lack of use as a key problem in OGD development [6–9]. Thus, in order to stimulate the utilization of OGD, many studies have been carried out to understand decisive factors of OGD usage [6, 7]. However, reviews have shown that these studies dig into this issue from aspects of either portals or users, but failed to consider the interaction between the two [10]. Moreover, "users of OGD are relatively less researched as subjects" in the present literature [8, p. 16]. Therefore, we could recognize a gap in present studies for analyzing the interaction between portals and users from the aspect of users' data acquisition process, which puts users of OGD portals at a central position.

Since users are consumers of OGD utilization effects [8] including transparency in policy [11], participation in public administrations [1, 12], as well as economic benefits [13], investigating their interaction with OGD portals could, therefore, benefit the future design of portals [14]. A better design of OGD portals based on a further understanding of OGD users could, then, help improve the efficiency of OGD utilization, and promote the positive results drawing from utilization [12].

In order to fill the identified research gap, we carried out an analysis of user's interaction with OGD portals during their data acquisition process based on Technology Acceptance Model (TAM). The main contribution of this paper is twofold: Firstly, to get a clear view of user's usage habits of OGD portals during the data acquisition process. Secondly, to evaluate the effect of user's characteristics on their interaction behavior. Based on the results, we also hope to find practical suggestions for future design of OGD portal interfaces to improve user's utilization of OGD.

The structure of the remaining paper is as follows. Section 2 presents the theoretical foundation of the research design. Section 3 explains in detail about the method of our analysis. We present our results in Sect. 4, followed by discussions of the results in Sect. 5. The paper ends with final conclusions in Sect. 6.

2 Theoretical Foundation

To form the foundation for our study, as well as to support the design of a survey instrument, we have drawn on related research carried out by both scholars and organizations related to OGD portals, OGD users and TAM.

2.1 Open Government Data Portal

Open government data is usually treated as a combination of two concepts: open data and government data [15]. The definition of "open data" given by Open Knowledge International is commonly accepted by scholars, which are data "that can be freely used, modified, and shared by anyone for any purpose" [16]. While government data are data, including products and services, "generated, created, collected, processed, preserved, maintained, disseminated, or funded by or for a government or public institution" [17, p. 4]. OGD portals are official portals launched by different levels of government to make governmental datasets publicly available to the end-users [2, 3, 5, 18], such as data.gov of the U.S. Government, data.gov.uk of the United Kingdom, and data.gov.sg of Singapore, etc. In this study, we treated OGD portal as a "one-stop-shop" for users to get access to OGD.

2.2   Users of Open Government Data Portal

Users are the main actor in OGD utilization process due to their direct effect on the coordination of utilization from the demand side [9]. Therefore, many studies emphasize user participation in extracting the value of OGD [8]. In the scope of human-computer interaction (HCI), user-centered design is one of the key disciplines, focusing on making the system as easy and pleasant to use as possible [19]. Usually, scholars would divide users of OGD into different types [20], including citizens [21, 22], business [23, 24], researchers [25, 26], developers [27], and journalists [4]. We note intersections between these user types because some of them are broad concepts (like citizens), while others are more specific concepts (like researchers and journalists).

2.3   Technology Acceptance Model

TAM is one of the most widely used theories of technology adoption [28]. It has been tested and validated for various kinds of users and systems, including e-government services [29]. The main aim of TAM is to explain and predict users' acceptance, thus "user" is a core concept in TAM [28]. According to TAM, user's attitude towards the utilization of a system is influenced by perceived usefulness (PU) and perceived ease of use (PEOU) [30]. Davis [30, p. 320] has defined PU as "the degree to which a person believes that using a particular system would enhance his or her job performance", and PEOU as "the degree to which a person believes that using a particular system would be free of effort". PEOU could also influence PU, because users would benefit more from a system if it is easier to use.

3   Research Design and Methods

The primary aim of this study is to find out user's usage habits in their interaction with OGD portals during the data acquisition process. To reach this aim, we built a research model based on TAM including three aspects: data acquisition methods, data quality (i.e. non-functional) requirements, and helping functions (Fig. 1). These three aspects cover users' data acquisition process from an OGD portal.

In data acquisition, we have included four different methods of finding data on a portal, namely keyword search, browse, ranking, and recommendation. We selected these methods based on prior visits to more than 10 different OGD portals and found that these were the possible methods users could use to discover the data they needed from a portal. "Keyword search" means entering queries to get the matching list of data. "Browse" means looking through records of data gathered according to certain rules like subject, government department, updated time, etc. "Ranking" means the top ranks of data of a certain category, like the most clicked datasets rank, the most downloaded datasets rank, etc. "Recommendation" includes two kinds: one is recommending similar or related data/datasets of the dataset that the user is looking at, which we refer to as *related*, the other is recommending datasets that users may have interests in according to their visiting history, which we refer to as *regular*.

We included data quality in the analysis, which is commonly perceived by scholars as the fitness for use by data consumers [1, 31, 32]. We find the fitness for use concept to be closely related with the definition of PU in TAM [30], as well as the core issue in HCI, which is usability [19]. As listed in Table 1, we collected 6 elements reflecting data quality from principles for OGD development [33–37].

The third aspect is helping functions, which indicates help offered by OGD portals to guide and assist users to find and use data or datasets on the portal. We selected four helping functions offered by OGD portals based on our prior visits to those portals. These functions are: user guide webpage, FAQ webpage, online smart agent to offer instant conversation, and customer phone.

According to the definition of PU and PEOU in TAM combining with our research model, we have proposed four hypotheses:

H1: Data acquisition methods relate to PEOU of users.
H2: Data acquisition methods influence user's needs of different data qualities.
H3: User's demands for helping functions relate to PEOU of users.
H4: User's demands for data quality attributes relate with their demands of helping functions.

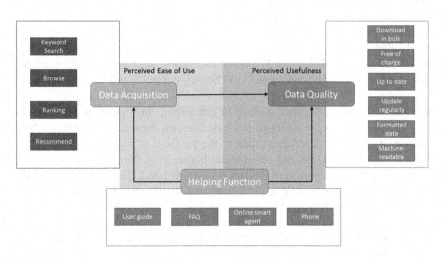

Fig. 1. Research model

A survey was designed and developed by operationalizing questions that reflect the aims and hypotheses of our study. The survey includes four parts. In the first part, there are several questions about the socio-demographic characteristics of users including gender, age, educational background, and occupation. The second part contains questions relating to data acquisition methods. We asked users about their acceptance of a certain method, their usage frequency and preferences of that method. A 7-point

Table 1. Description for elements of data quality

Element	Description	Reference
Download in bulk	Possibility to download all of the data needed at once	[33, 34]
Formatted data	Data is provided with certain structure including meta-data information	[33, 35]
Free of charge	Don't need to pay for the data/dataset	[34, 36, 37]
Machine-readable	Data/dataset is stored in widely-used file formats that easily lend themselves to machine processing, accessing and modifying single elements in the file	[33–35]
Up to date	Data/dataset is made available as quickly as necessary to preserve the value of the data	[35, 37]
Update regularly	To maintain the sustainable utilization of a data/dataset the data/dataset is kept regularly updated	[33, 37]

Likert scale (from "Strongly don't like" to "Strongly like", or "Strongly not frequent" to "Strongly frequent") is used for exploring users' preferences and usage frequency of different data acquisition methods. This part ended with the question asking user's feeling of difficulty when using the portal. Users could rank their feelings from 1 (Strongly not difficult) to 7 (Strongly difficult). The third part includes questions asking about users' need of certain data qualities by using a 7-point Likert scale (from "Strongly don't need" to "Strongly need"). The final part is questions relating to helping functions of OGD portals. We asked users about their need of help when using the portal (from "Strongly don't need" to "Strongly need"), and their preferences of helping types.

We used an online tool called *Sojump* to create the online questionnaire. We chose citizens to represent users of OGD portals based on two reasons. On one hand, the key motivation of releasing government data to the public is to reduce the asymmetry of information between citizens and government bodies [38], thus citizens are important stakeholders in the development of OGD. On the other hand, choosing citizens as study objects offers the advantages of wider representation of the population and range of diverse characteristics [9]. Due to the recent efforts of the Chinese government in developing OGD portals, we distributed our online questionnaire in a sample of the Chinese citizens from August 1st to 10th 2017 through WeChat, E-mails, and Weibo. Because there was no direct motivation for citizens in China to engage in such a survey, we had to apply convenience sampling. Although this is not an ideal method of collecting research data, it is not uncommon in scientific research [12]. This is because the increasing non-response rates in recent years have caused samples improper for statistical inferences [39], and efforts against non-response are too complex and costly but may still be useless [40]. Thus, we chose to continue the data analysis with the convenience sample.

4   Results

In total, we have received 208 valid responses. Table 2 shows the socio-demographic characteristics of the responses. Females represented 63% of all respondents, and males represented 36.5%. Our samples covered a wide range of ages and were almost evenly distributed in age groups 26–30, 31–40, and 41–50, which were also the main groups with a total percentage of 81.8%. Referring to the education background, most respondents were in the group of undergraduate (52.9%). Our samples covered all kinds of occupations listed in the survey, with student, manager and teacher being the top three popular ones.

Table 2. Socio-demographic characteristics of the sample

Topic	Dimension	Frequency	Percent	Topic	Dimension	Frequency	Percent
Gender	Male	76	36.5%	Occupation	Student	24	115%
	Female	131	63.0%		Production worker	9	4.3%
	Other	1	0.5%		Marketing/Salesman	18	8.7%
Age	Under 18	2	1.0%		Customer service	4	1.9%
	18–25	10	4.8%		Logistics	16	7.7%
	26–30	54	26.0%		Human resources	7	3.4%
	31–40	57	27.4%		Financial/auditor	12	5.8%
	41–50	59	28.4%		Civilian past	12	5.8%
	51–60	24	11.5%		Technician	15	7.2%
	Over 60	2	1.0%		Manager	33	15.9%
Educational qualification	Junior high	1	0.5%		Teacher	22	10.6%
	Senior high	28	13.5%		Consultant	4	1.9%
	Under graduate	110	52.9%		Specialist	9	4.3%
	Post graduate	57	27.4%		Other	23	11.1%
	Beyond post graduate	12	5.8%	Total		208	100.0%

We have examined the reliability of all the scales in our survey with Cronbach's alpha [41]. The commonly accepted range for alpha is 0.70 to 0.95 [42]. Our results therefore show high reliability with a score of 0.864 (Table 3). We also examined all the scales in the survey with Kaiser-Meyer-Olkin (KMO) measure of sampling adequacy [43]. The result is 0.923, which indicates the variables to be suitable for factor analysis [44]. The significance of Bartlett's test of sphericity is less than 0.05, which also indicated high validity of the scales. Thus, we could believe the survey for analyzing user's interaction with the portal during data acquisition process to be reliable.

Since our primary goal of this study is to find out user's usage habits in their interaction with OGD portals during the data acquisition process, we first carried out data analysis from three aspects: data acquisition methods, data quality requirements, and helping functions. We also combined the results with respondents' socio-demographic characteristics to find out whether there are any relations among these variables.

Table 3. Reliability and adequacy test of the survey

Variable No.	Valid	%	Cronbach's Alpha
12	162	78%	0.864
Bartlett's test of sphericity			Kaiser-Meyer-Olkin measure of sampling adequacy
Approx. Chi-square df	df	Sig.	
1240.442	66	0.000	0.807

4.1   Users' Data Acquisition Method

Generally, users of OGD portals felt it quite difficult to get the data they need from the portal (Fig. 2). Since the sample size for user's socio-demographic characteristics are the same with their feeling of difficulty, it is statistically robust [45] to carry out two independent samples T-test to determine the relationship between their feeling of difficulty and their gender, as well as one-way Anova test of users' feeling of difficulty with their age, education background and occupation. Results show that only educational certification has a significant effect on people's feeling about data access difficulty (F = 2.95, P = 0.021).

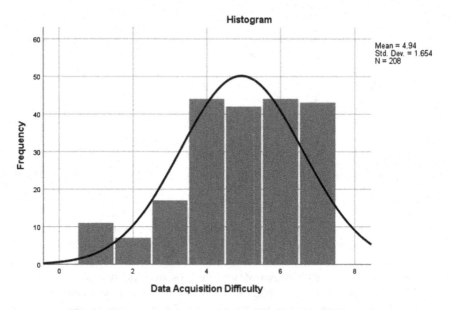

Fig. 2. Histogram of data acquisition difficulty with OGD portals

Figure 3 shows results regarding acceptance of the four methods of data acquisition, namely keyword search, browse, ranking, and recommendation. We found over 90% of the respondents use keyword search for the data they wanted on a portal, followed by the related datasets automatically recommended by the portal. About 25.5% of the respondents chose not to refer to the rankings when looking for datasets on a portal, making it the method with the lowest acceptance rate.

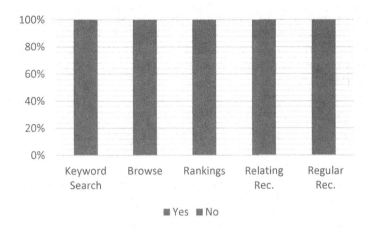

Fig. 3. Users' acceptance of data acquisition methods

We compared respondents' reported usage frequency and preferences for keyword search and browsing (Fig. 4). According to the results, we found that more respondents chose to use keyword search to find the data they need, since 190 out of 208 respondents have ever used keyword search, while only 170 used browsing. However, the mean score for the usage frequency of browsing was much higher than keyword search, with a relatedly small standard deviation. This indicated that using frequency for browsing resulted in a higher value than that of keyword search. On the other hand, keyword search got a higher preference score than browsing.

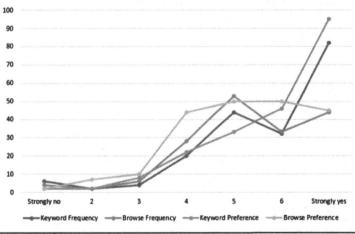

	Keyword Frequency	Browse Frequency	Keyword Preference	Browse Preference
Valid N.	190	170	208	208
Mean	5.73	5.88	5.35	5.23
Std. Deviation	1.472	1.332	1.381	1.387

Fig. 4. Reported usage frequency & preferences of keyword search and browse

We have carried out chi-square test of whether gender influences people's acceptances of regular recommendation of datasets based on their visiting history. The likelihood ratio (=10.501) shows a strong relationship between the two (Sig (2-sided) p = 0.005). Compared with males, females are more likely to accept regular recommendations (Fig. 5).

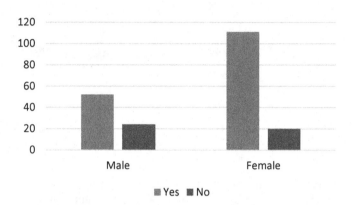

Fig. 5. Differences of acceptance of regular recommendations between genders

4.2   Users' Need of Data Quality

We first calculated respondents' needs for different types of data quality on OGD portals. Results have shown that *up to date* got the highest mean value and lowest standard deviation, while *download in bulk* got the lowest score but highest standard deviation (Fig. 6).

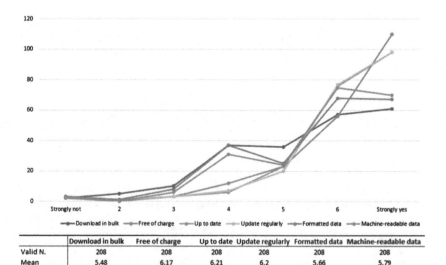

	Download in bulk	Free of charge	Up to date	Update regularly	Formatted data	Machine-readable data
Valid N.	208	208	208	208	208	208
Mean	5.48	6.17	6.21	6.2	5.66	5.79
Std. Deviation	1.404	1.19	1.014	1.056	1.309	1.229

Fig. 6. Users' need of different data quality

We carried out one-way Anova to analyze the effect of age, education and occupation on respondents' demand of different data qualities (Table 4). We found that age has a significant effect on *download in bulk*, *up to date* and *formatted data*. People of different educational background have significant different demands for free data, up to date data, and data being updated regularly. Occupation affected the demand for *download in bulk*, *formatted data* and *machine-readable data*.

Table 4. One-way Anova of age, education, occupation and user's needs of data quality

Group by	Data type	F	P	Levene	P
Age	Download in bulk	6.276	0.000	2.148	0.05
	Up to date	2.539	0.022	0.649	0.691
	Formatted data	2.817	0.012	0.919	0.482
Educational background	Free of charge	3.493	0.009	2.632	0.051
	Up to date	3.007	0.019	0.443	0.723
	Update regularly	2.637	0.035	1.225	0.302
Occupation	Download in bulk	1.891	0.033	1.380	0.172
	Formatted data	3.645	0.000	1.736	0.056
	Machine-readable	2.417	0.005	1.512	0.116

4.3   Helping Functions

We first calculated the percentage of respondents' acceptance of helping functions of OGD portals. 90.9% of the respondents needed help from the portal. The Chi-square tests of people's acceptance of helping functions of a portal show no relation to their gender, age, occupation or educational background. Referring to users' extent of need of helping functions, results show users' strong demand for OGD portals to offer help, with the mean of 5.34, and the standard deviation of 1.331. Their choices gathered in 5 (26.4%) and 7 (25%) of the 7-point scale.

We also calculated the frequencies and percentages of users' choices of different helping functions (Fig. 7). The differences are not significant among different kinds of helping functions. FAQ is preferred by most respondents, while customer service

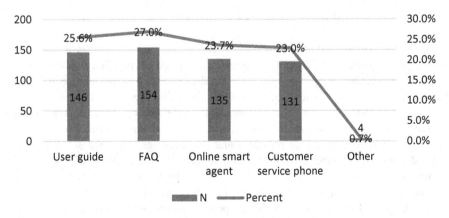

Fig. 7. Frequencies and percentage of helping functions

phone receives least selections. The Chi-square tests of people's choices of helping functions show no relation to their gender, age, occupation or education background.

5 Discussion

Based on the results of the survey, we further carried out analysis of the relationships among data acquisition methods, data quality requirements and helping functions.

We first examined the relation of PEOU and users' methods of data acquisition. There were no significant differences of the means of users' feeling of difficulties among different data acquisition methods (Fig. 8). However, we noticed a great gap in the mean between users using keyword search and those who don't, although the two independent samples T-test shows the differences between these two groups are not significant. Thus, we could conclude that, generally, users of OGD portals found it difficult to find the data they needed from the portal, but there were no significant differences of PEOU among different data acquisition methods. Thus, our first hypothesis (H1) was not supported by the data.

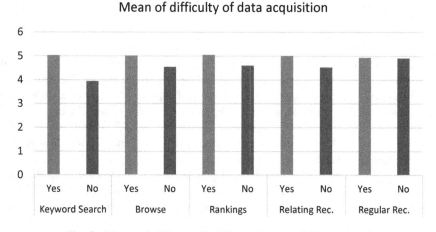

Fig. 8. Means of difficulty for different data acquisition methods

Secondly, we analyzed the relation between users' data acquisition methods and their needs for different data quality attributes. We carried out two independent samples T-test to see whether users' need for different data quality has any relation with their data acquisition method. The results (Table 5) show that users who browse seek data that is *free of charge*, *up to date*, and *machine-readable* more than users who don't browse. Users who refer to the ranks provided are in greater need of free data than users who don't. Users who accept recommendations show greater need for data to be updated regularly. On the other hand, we examined the correlations of users' usage frequency and preferences of keyword search and browsing with their demands for different data qualities. According to the Pearson correlation scores in Table 6, we found that generally, there were significant positive correlations between users' need of

Table 5. T-test of data acquisition methods and data quality

		Free of charge						Up to date						Update regularly						Machine-readable					
		Mean	Std.D	F	P	t	P	Mean	Std.D	F	P	t	P	Mean	Std.D	F	P	t	P	Mean	Std.D	F	P	t	P
Browse	Yes	6.28	1.06	20.26	0.00	2.32	0.03	6.31	0.90	9.66	0.00	2.40	0.02	–	–	–	–	–	–	5.88	1.16	5.58	0.02	2.05	0.05
	No	5.66	1.58					5.76	1.34					–	–					5.37	1.44				
Rankings	Yes	6.34	1.05	8.67	0.00	3.20	0.00							6.34	0.92	9.35	0.00	3.21	0.00	–	–				
	No	5.66	1.43											5.85	1.18					–	–				
Relating Rec.	Yes	–	–					–	–				–	6.33	0.89	17.03	0.00	3.28	0.00	–	–				–
	No	–	–					–	–					5.40	1.52					–	–				
Regular Rec.	Yes	–	–					–	–				–	6.34	0.91	7.92	0.01	3.14	0.00	–	–				–
	No	–	–					–	–					5.67	1.37					–	–				

Table 6. Pearson correlation analysis of data acquisition and data quality

		Download in bulk		Free of charge		Up to date		Update regularly		Formatted data		Machine-readable	
		r	P	r	P	r	P	r	P	r	P	r	P
Keyword search	Frequency	0.309	0.000	0.371	0.000	0.370	0.000	0.368	0.000	0.306	0.000	0.292	0.000
	Preference	0.430	0.000	0.530	0.000	0.505	0.000	0.511	0.000	0.410	0.000	0.425	0.000
Browse	Frequency	0.202	0.008	0.249	0.001	0.308	0.000	0.326	0.000	0.257	0.001	0.321	0.000
	Preference	0.240	0.000	0.349	0.000	0.351	0.000	0.352	0.000	0.284	0.000	0.311	0.000

*r stands for Pearson Correlation, P stands for 2-tailed significant

data qualities and their preferences and usage frequency of keyword search and browsing. Moreover, correlations are stronger with their preferences of data acquisition methods, than with their usage frequencies. The above results supported our second hypothesis (H2).

Thirdly, we examined the correlations of users' need of help with their feeling of difficulty of data acquisition from OGD portals. Results have shown that there was a strong possibility (Sig.(2-tailed) p = 0.001) that these two have a positive relation, with Pearson Correlation being 0.236. The result shows support for our third hypothesis (H3) that user's demands of helping functions relates to the PEOU of users.

Finally, we analyzed the relation between helping functions and users' demand of data qualities. The Pearson correlation scores in Table 7 show that users' needs for help has a significant positive relation with their needs for all the data quality attributes considered. The two independent samples T-test of the relation between different helping functions and users' needs for different data qualities shows that users who choose to use FAQ have a relatively higher demand of data qualities (Table 8). The relation is not significant with other helping functions and users' need of data qualities. Thus, the above results show support for our fourth hypothesis (H4) that user's demands of data qualities have certain relationship with their demands of helping functions.

Table 7. Pearson correlation analysis of need of help and data quality

Download in bulk	Free of charge	Up to date	Update regularly	Formatted data	Machine-readable
r = 0.399*	r = 0.452*	r = 0.433*	r = 0.443*	r = 0.405*	r = 0.405*

r = Pearson Correlation. 2-tailed, *significance p < 0.05

Table 8. T-test of FAQ and data quality

		Free of Charge						Up to date						Update Regularly					
		Mean	Std.D	F	P	t	P	Mean	Std.D	F	P	t	P	Mean	Std.D	F	P	t	P
FAQ	Yes	6.330	0.997	7.928	0.005	-2.800	0.007	6.380	0.777	15.345	0.002	-3.311	0.002	6.380	0.856	9.941	0.002	-3.476	0.001
	No	5.700	1.537					5.720	1.371					5.690	1.371				

		Formatted Data						Machine-readable					
		Mean	Std.D	F	P	t	P	Mean	Std.D	F	P	t	P
FAQ	Yes	5.880	1.157	7.325	0.007	-3.614	0.001	6.000	1.073	10.722	0.001	-3.801	0.000
	No	5.060	1.522					5.190	1.442				

6 Conclusion

OGD portals are the most commonly implemented approaches for the publishing and consuming of open government data for citizens [1], and users of OGD portals are consumers of OGD [8] who have a direct effect on OGD utilization [9], thus looking into the interaction between users and OGD portals could help improve OGD utilization as well as benefits derived from its usage.

In order to analyze the interaction between OGD users and portals, we carried out this study from the perspective of citizens representing OGD users which covered three aspects including data acquisition methods, data quality requirements, and helping functions, as well as the relations among these three. We designed and carried out an online survey among Chinese citizens to collect the data for analyzing, and drew the following key conclusions:

- Users of OGD portals accept keyword search most in four data acquisition methods but browsing has the highest and most stable usage frequency.
- Females are more likely to accept regular recommendations of data/datasets than males.
- Users' age, education background and occupation affect their demands for different data qualities.
- Users' demands for data quality attributes are positively related to their preferences and usage frequency of data acquisition methods.
- Users' needs for help are positively related with their feelings of difficulty to get data from OGD portals.
- Users' needs for help are positively related to their demands for different data qualities.

Based on the above conclusions drawn from the data analysis, we recommend future development of OGD portals to offer better helping functions to reduce data acquisition difficulties. Improving data qualities, especially keeping data on the portal up to date, could also help reduce the difficulties for data access from the portal.

Our study has both strengths and limitations. Based on TAM, we have analyzed users' habits of interacting with OGD portals. Our conclusion also accords with TAM in PEOU's influencing on PU. Our methodology, however, has some limitations. Firstly, our survey respondents only include a sample from the Chinese population. Thus, the results may not be generalizable to other populations in other countries. On the other hand, since the three elements in our research model (data acquisition, data quality and helping functions) are not specific to government portals, the model could potentially be used to evaluate the knowledge acquisition process in non-government open data portals. Evaluating the generality and applicability of the model for other types of data portals and populations is left as future research. As a second limitation, the use of convenience sampling may have an effect on the reliability of the analysis. But as explained in Sect. 3, it was the most appropriate choice for this study. In the future, we plan to find out specific methods to improve the helping functions of OGD portals, so that users could more easily find and use the data on OGD portals.

References

1. Attard, J., et al.: A systematic review of open government data initiatives. Gov. Inf. Q. **32**(4), 399–418 (2015)
2. Lourenço, R.P.: Open government portals assessment: a transparency for accountability perspective. In: Wimmer, M.A., Janssen, M., Scholl, H.J. (eds.) EGOV 2013. LNCS, vol. 8074, pp. 62–74. Springer, Heidelberg (2013). https://doi.org/10.1007/978-3-642-40358-3_6
3. Lourenço, R.P.: An analysis of open government portals: a perspective of transparency for accountability. Gov. Inf. Q. **32**(3), 323–332 (2015)
4. Heise, A., Naumann, F.: Integrating open government data with stratosphere for more transparency. Web Semant. Sci. Serv. Agents World Wide Web **14**, 45–56 (2012)
5. Kassen, M.: A promising phenomenon of open data: a case study of the Chicago open data project. Gov. Inf. Q. **30**(4), 508–513 (2013)
6. Ruijer, E., et al.: Connecting societal issues, users and data. Scenario-based design of open data platforms. Gov. Inf. Q. **34**, 470–480 (2017)
7. Wang, H.-J., Lo, J.: Adoption of open government data among government agencies. Gov. Inf. Q. **33**(1), 80–88 (2016)
8. Safarov, I., Meijer, A., Grimmelikhuijsen, S.: Utilization of open government data: a systematic literature review of types, conditions, effects and users. Inf. Polity **22**, 1–24 (2017)
9. Zuiderwijk, A., Janssen, M.: A coordination theory perspective to improve the use of open data in policy-making. In: Wimmer, M.A., Janssen, M., Scholl, H.J. (eds.) EGOV 2013. LNCS, vol. 8074, pp. 38–49. Springer, Heidelberg (2013). https://doi.org/10.1007/978-3-642-40358-3_4
10. Meijer, A., de Hoog, J., van Twist, M., van der Steen, M., Scherpenisse, J.: Understanding the dynamics of open data: from sweeping statements to complex contextual interactions. In: Gascó-Hernández, M. (ed.) Open Government. PAIT, vol. 4, pp. 101–114. Springer, New York (2014). https://doi.org/10.1007/978-1-4614-9563-5_7
11. Florini, A.: Making transparency work. Glob. Environ. Politics **8**(2), 14–16 (2008)
12. Wijnhoven, F., Ehrenhard, M., Kuhn, J.: Open government objectives and participation motivations. Gov. Inf. Q. **32**(1), 30–42 (2015)
13. Willinsky, J.: The unacknowledged convergence of open source, open access, and open science. First Monday **10**(8) (2005)
14. Galitz, W.O.: The Essential Guide to User Interface Design: An Introduction to GUI Design Principles and Techniques. Wiley, New York (2007)
15. Ubaldi, B., Open government data: towards empirical analysis of open government data initiatives. In: OECD Working Papers on Public Governance, vol. 22, p. 1 (2013)
16. The Open Difinition. https://opendefinition.org/. Accessed 9 June 2018
17. OECD: Recommendation of the Council for enhanced access and more effective use of Public Sector Information (2008)
18. Kostovski, M., Jovanovik, M., Trajanov, D.: Open data portal based on semantic web technologies. In: Proceedings of the 7th South East European Doctoral Student Conference. University of Sheffield, Greece (2012)
19. Dix, A.: Human-computer interaction. In: Liu, L., Özsu, M.T. (eds.) Encyclopedia of Database Systems, pp. 1327–1331. Springer, Boston (2009). https://doi.org/10.1007/978-0-387-39940-9_192
20. King, W.R., He, J.: A meta-analysis of the technology acceptance model. Inf. Manag. **43**(6), 740–755 (2006)

21. Parycek, P., Hochtl, J., Ginner, M.: Open government data implementation evaluation. J. Theor. Appl. Electron. Commer. Res. **9**(2), 80–99 (2014)
22. Power, R., Robinson, B., Rudd, L., Reeson, A.: Scenario planning case studies using open government data. In: Denzer, R., Argent, R.M., Schimak, G., Hřebíček, J. (eds.) Environmental Software Systems, Infrastructures, Services and Applications, ISESS 2015. IFIP Advances in Information and Communication Technology, vol. 448, pp. 207–216. Springer, Cham (2015). https://doi.org/10.1007/978-3-319-15994-2_20
23. Magalhaes, G., Roseira, C., Manley, L.: Business models for open government data. In: Proceedings of the 8th International Conference on Theory and Practice of Electronic Governance. ACM (2014)
24. Susha, I., Grönlund, Å., Janssen, M.: Driving factors of service innovation using open government data: An exploratory study of entrepreneurs in two countries. Inf. Polity **20**(1), 19–34 (2015)
25. Gonzalez-Zapata, F., Heeks, R.: The multiple meanings of open government data: understanding different stakeholders and their perspectives. Gov. Inf. Q. **32**(4), 441–452 (2015)
26. Whitmore, A.: Using open government data to predict war: a case study of data and systems challenges. Gov. Inf. Q. **31**(4), 622–630 (2014)
27. Veeckman, C., van der Graaf, S.: The city as living laboratory: empowering citizens with the citadel toolkit. Technol. Innov. Manag. Rev. **5**(3) (2015)
28. Venkatesh, V., Davis, F.D.: A theoretical extension of the technology acceptance model: four longitudinal field studies. Manag. Sci. **46**(2), 186–204 (2000)
29. Carter, L., Bélanger, F.: The utilization of e-government services: citizen trust, innovation and acceptance factors. Inf. Syst. J. **15**(1), 5–25 (2005)
30. Davis, F.D.: Perceived usefulness, perceived ease of use, and user acceptance of information technology. MIS Q., 319–340 (1989)
31. Dawes, S.S.: Stewardship and usefulness: policy principles for information-based transparency. Gov. Inf. Q. **27**(4), 377–383 (2010)
32. Wang, R.Y., Strong, D.M.: Beyond accuracy: what data quality means to data consumers. J. Manag. Inf. Syst. **12**(4), 5–33 (1996)
33. Ten Principles for Opening Up Government Information. https://sunlightfoundation.com/policy/documents/ten-open-data-principles/. Accessed 9 June 2018
34. Methodology - Global Open Data Index. https://index.okfn.org/methodology/. Accessed 9 June 2018
35. 8 Principles of Open Government Data. https://public.resource.org/8_principles.html. Accessed 9 June 2018
36. Board, P.S.T. (ed.): Public Data Principles (2012)
37. OpenDataBarometer ODB Methodology - v1.0 (2016)
38. Murillo, M.J.: Evaluating the role of online data availability: the case of economic and institutional transparency in sixteen Latin American nations. Int. Polit. Sci. Rev. **36**(1), 42–59 (2015)
39. Peytchev, A.: Consequences of survey nonresponse. ANNALS Am. Acad. Polit. Soc. Sci. **645**(1), 88–111 (2013)
40. Schmeets, H., Janssen, J.P.: Using national registrations to correct for selective nonresponse. Political preference of ethnic groups. Statistics Netherlands (2003)
41. Cronbach, L.J.: Coefficient alpha and the internal structure of tests. psychometrika **16**(3), 297–334 (1951)
42. Tavakol, M., Dennick, R.: Making sense of Cronbach's alpha. Int. J. Med. Educ. **2**, 53 (2011)

43. Kaiser, H.F.: A second generation little jiffy. Psychometrika **35**(4), 401–415 (1970)
44. Dziuban, C.D., Shirkey, E.C.: When is a correlation matrix appropriate for factor analysis? Some decision rules. Psychol. Bull. **81**(6), 358 (1974)
45. Posten, H.O.: The robustness of the two—sample t—test over the Pearson system. J. Stat. Comput. Simul. **6**(3–4), 295–311 (1978)

Blockchain: Trends and Future

Wenli Yang, Saurabh Garg$^{(\boxtimes)}$, Ali Raza, David Herbert, and Byeong Kang

Discipline of ICT, School of TED, University of Tasmania, Sandy Bay, Australia
{yang.wenli,saurabh.garg,ali.raza,david.herbert,byeong.kang}@utas.edu.au

Abstract. Blockchain has attracted a great deal of attention due to its secure way of distributing transactions between different nodes without a trust entity, and tracking the validity of data. Although many experts argue the solutions to several problems in today's inherently insecure Internet lies with blockchain technology because of its security and privacy features, there is no systemic survey to analyze and summarize blockchain technology from different perspectives. In this paper, we present the current trends in blockchain technology from both technical and application viewpoints and highlight the key challenges and future work required that will help in determining what is possible when blockchain is applied to existing and future problems.

Keywords: Blockchain · Scalability · Crypto-currency

1 Introduction

For the last few decades there has been massive volumes of digital information produced due to the growth of computing technologies such as storage, processing power and networking. On one side, we can see how the maintenance of data has been transformed from purely private information on isolated desktops to completely public data in the form of social networks. On the other side, every IT service is becoming outsourced to third parties in the form of Cloud computing [4]. Moreover, we can see the digital data growth has been so enormous that it is called "Big Data" and it brings more opportunities to innovate and optimize our decisions [13]. However, such growth has also led to grave issues in terms of trust, privacy and security [6]. The problems such as 'Fake News' are also coming into focus [1]. The recent third-party distribution of millions of Facebook user's data has compounded the problem further and it increased public awareness of the issues. Given all of our data is public and maintained in a decentralised manner, it is almost impossible to keep track of such issues. In the real world we find the installment of CCTV cameras to trace criminal activity has not only reduced crime but it has also put fear in the mind of criminals that are being watched. We postulate the question as to whether we have something analogous to CCTV cameras for all of our internet activities that can give some sense of protection for of our data.

Recent development and applications of distributed ledgers such as blockchain has given a glimpse as to how to alleviate such issues [19]. One of the

K. Yoshida and M. Lee (Eds.): PKAW 2018, LNAI 11016, pp. 201–210, 2018.
https://doi.org/10.1007/978-3-319-97289-3_15

key examples is Bitcoin, which is a decentralized currency that is maintained in an autonomous manner. All transactions processed in Bitcoin are tracked through a "blockchain", where each transaction is verified in a decentralised manner before it is recorded, preventing any chance of illegitimate transactions occurring. Due to the potential use of blockchain and related technologies to maintain tamper-proof systems that can be maintained in a distributed and autonomous manner, blockchain seems to be the perfect match to make various sectors that operate through a public network such as the Internet accountable. With this aim in mind, this paper will review the various trends in blockchain to understand how it can play an important role in protecting digital data and its usage. Based on the review, we also give a roadmap for future research that is required to make this vision possible.

In summary, the contributions of our research are:

- We provide a detailed evolution of blockchain systems, and discuss technical resources and typical characteristics of different stages of development.
- We describe and categorize trends in blockchain from three criteria: data structure, consensus method and the overall system.
- We introduce several blockchain platforms and compare them using criteria related to usability, limitation, flexibility and performance, and this summary can serve to guide the future blockchain research and development.
- We discuss future challenges in the design of blockchain-based Internet ecosystems, and how the application of AI can guide future implementations.

This paper is organized as following: Sect. 2 introduces the basic concepts of blockchain. Section 3 outlines the existing data structures used and Sect. 4 describes blockchain's most common consensus algorithms. In Sect. 5 we discuss trends in blockchain systems.

2   Blockchain Basics

To understand blockchain, we need to understand the meaning of a distributed ledger. A typical distributed ledger is a shared database that is replicated, and synchronized in a decentralised manner among the different members of a network. The distributed ledger stores the transactional data of the participants in the network. A blockchain is based on Distributed Ledger Technology (DLT) that is spread across several nodes or computing devices [14].

It is assumed that these nodes do not fully trust each other as some may exhibit Byzantine (dishonest) behaviour. These nodes maintain a long chain of cryptographic hash-linked blocks where each modification or addition of a transaction is validated by the consensus of all nodes in the system. In one sense blockchain is similar to a traditional database requiring ACID properties to be satisfied. The key difference is the 'distributed consensus' algorithm which decides whether a new block is valid and legitimate before any insertion can be done. Based on the membership of the nodes, the blockchain can be either public, private or hybrid [2]. Public blockchains are fully decentralized where any node can join and leave the system. The other two types enforce some restrictions on membership in regard to system access. Hybrid blockchains tries to combine the characteristics of both private and public blockchains – every transaction is private, however it can be verified in the public state. Figure 1 shows the data structure of a blockchain whose basic concepts include:

1. **Transaction:** an operation that caused a change of the block.
2. **Block:** a container data structure, and a block is composed of a header and a long list of transactions.
3. **Chain:** is a continuously growing list of blocks, which are linked and secured using cryptography.

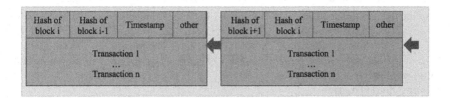

Fig. 1. Blockchain datastructure

3   Trends in Blockchain Type Data Structure

As stated before, blockchain technology originated from Distributed Ledger Technology (DLT) proposed in the 1990s [18]. Even though DLT was proposed almost 20 years ago, it became prominent with the implementation of Bitcoin and its usage as a crypto-currency in 2008 [14]. The initial data structure for blockchain was based on a hash-table, however with its significant growth of usage, it became apparent that new data structures are required for the efficient storage and transmission of information in order to maintain a rapidly growing number of transactions in the ledger. Due to these requirements, several new data structures were proposed for solving the limitations of the traditional blockchain. Some authors [9] suggested the usage of Directed Acyclic Graph (DAG) for maintaining transaction information as it is scalable, lightweight and decentralised. However, this alternative also has similar problems to blockchain at very high scales. Tempo Ledger proposed by RadixDLT aims to scale linearly in an unbounded and efficient manner [16]. In the Tempo ledger, each node can maintain a shard of the whole ledger in comparison to the traditional blockchain implementation where the whole ledger is maintained at each node. We summarised these trends from different perspectives in Table 1.

Table 1. Trends in blockchain type data structure

Year	1995	2008	2017	2018
DS	MDL	Blockchain	Directed Acyclic Graph or DAG	Tempo ledger
Applications	Sharing Economy Broker and Reinsurance Claim Payment Data Sharing	Cyrptocurrencies e.g. Bitcoin, Litecoin, Ripple, Namecoin, etc.	Iota, Raiblocks, Hashgraph, etc.	RadixDLT
TF	High	High	Low	Low
TCT	Several minutes	Several minutes	Minutes	<5 s
Popularity	Launched in 1995, used by a few systems now	Launched in 2008. Very well known	Launched in 2017. Not well known yet	Will be Launched in Q3 2018

MDL = *Mutual distributed ledger*: a record of transactions shared in common and stored in multiple locations.
DS = key data structure
TF = Transaction fee
TCT = Transaction confirmation time

4 Trends in Consensus Algorithms

Consensus algorithms are designed to achieve reliability in a network involving multiple unreliable nodes. They ensure that the next block in a blockchain is the one and only version of the truth, and it keeps adversarial groups from derailing the system and successfully forking the chain. The most common consensus algorithms include Proof of Work (PoW), Proof of Stake (PoS) [11], Delegated Proof of Stake (DPoS) [11], Ripple [17], Practical Byzantine Fault Tolerance (PBFT) [5] and Delegated Byzantine Fault Tolerance (dBFT). A summary based on different application scenarios and features of the consensus mechanism by the following criteria is presented in Table 2.

- **Data management:** support for whole network nodes and data supervision by privilege nodes.
- **Performance and efficiency:** confirmed efficiency of consensus between transactions.
- **Resource consumption:** high CPU load, storage, network capacity, etc. during consensus processing.
- **Tolerance power:** anti-attacking and cheat-proof capacity.

The PoW protocol is one of the first utilised consensus protocols that is based on computational load, requiring *miners* to find a solution to a puzzle. Several cyrpto-currencies utilise a variant of this protocol. Performance is quite low and found to be not suitable for very large ecosystems. To reduce the high resource cost of mining, PoS was proposed that assigns a difficulty value to a puzzle based on how much *stake* the miner has in the network. Delegated Proof of Stake (DPoS) is a newer consensus structure where users select some delegate nodes that confirm the validity of a block. Some consensus protocols such dBFT and PBFT are based on communication between different nodes and they are mostly used in private chains having authenticated nodes. Tendermint [12] improves the performance of PBFT by making small modification allowing different nodes with different voting power. The voting power is determined by the stake a user owns in the network. Despite the many modifications to the original consensus protocol, they still fail to scale well. To overcome this, federated protocols such as Ripple were proposed. In these protocols, the whole network is partitioned into smaller units; and each unit runs a local consensus.

Table 2. Consensus algorithm summary

Consensus	PoW	PoS	DPoS	Ripple	Tendermint	PBFT	dBFT
Year	2008	2012	2014	2014	2014	2015	2016
Data management	O	O/P	O/P	O/P	P	P	P
Performance	L	M	M	H	H	H	H
High resource	Yes	Partial	Partial	No	No	No	No
Tolerance	<=25%	<=51%	<=51%	<=20%	<=33.3%	<=33.3%	<=33.3%
Application	Bitcoin	Tezos	Lisk	Ripple	Tenderminty	Hyperledger fabric	Neo

O = Open, P = Permission
L = Low, M = Medium, H = High

5   Trends in Blockchain Systems

Over the past few years, blockchain technology has been evolving rapidly – from the original Bitcoin protocol to the second generation Ethereum platform [15], and today we are in the process of building what is informally termed blockchain 3.0 and future-generational blockchain 4.0 (see Table 3). This evolutionary change shows how the technology is evolving from its initial form as essentially just a database, to becoming a fully-fledged globally distributed system. The applications of blockchain have evolved to much wider scopes than crypto-currency and asset management. Applications from different industries including healthcare and energy sectors are being designed with blockchain as an underlying technology. These application's requirements have also led to structural changes in blockchain itself, which is evolving from linear chains to DAG, with emerging future types of chains such as Relational and Divisible chains.

Blockchain 1.0 is completely dedicated to the decentralization of money and payments, although this was the first implementation of a distributed ledger technology (DLT). It supports the mining of Bitcoins. The network is peer to peer and transactions take place between users directly without the involvement of any third party. Other crypto-currencies that are recently supported are Litecoin, Dogecoin etc. The technology stack of bitcoin consists of the blockchain platform, and a protocol which is used to describe how assets are transferred. The concensus algorithm utilised is Proof of Work (PoW). Blockchain 1.0 guarantees distributed storage, enables data sharing between nodes, and enables transparency in transaction processing.

In Blockchain 2.0, a logic tier was added into the ledger and which supported what is termed smart contracts. Smart contracts are small computer programs that execute automatically when certain conditions are met. Since smart contracts are in essence tamper-proof, it reduces the cost of verification, execution and fraud prevention. The most prominent system in this version of blockchain is Ethereum. It is a platform for implementing smart contracts. It was proposed in 2013 and the initial release of its first blockchain was in July 2015. This version enables the creation and transfer of digital assets.

After the initial successes of Blockchain 1.0 and 2.0, several limitations were revealed. The most important ones are:

- **Energy consumption:** Since mining requires significant energy (electricity) costing billions of dollars per year, it is not scalable to mass adoption.

Table 3. Comparison of different generations of blockchain.

Evolution year	Blockchain 1.0 2008	Blockchain 2.0 2013	Blockchain 3.0 2015	Blockchain 4.0 2018
Apps	Digital currency	Smart contract	Decentralized applications (DApp)	Usable in wider industrial applications
CS	Meta chain	Meta chain	(a) Meta chain and side chain, (b) Directed graph data structure	(a) Relational chains (b) Divisible chains
SL	very limited	FFPL	FFPL	FFPL
Consensus	PoW	PoW, PoS	PoW, PoS, DPoS, PBFT, Ripple, PoET	Consensus algorithm based on AI
ID	Mining	Initial Coin Offering (ICO)	ICO, ZCASH, EOS	Seele and others
Features	guaranteed transaction authenticity; reduced server costs; transactions transparency	Guarantee of distributed computation; creating and transferring digital assets	Completely open-source; autonomous operation; arbitrary protocol language support	Faster consensus and transaction confirmation; complete ecosystem of bottom-up technologies and applications

CS = Chain Structure
SL = Scripting Language
ID = Initial Distribution FFPL = Fully Featured Programming Language

- **Volume of transactions:** The number of transactions is increasing every 10–12 s with each new block creation. Bitcoin can theoretically process 7 transactions per second while Ethereum processes 15 transactions per second. If we compare the number of transactions to Visa's network, which processes 24000 transactions per second, we still need to improve volume of transactions.
- **Cost:** Since a small fee is required to pay miners for maintaining the ledger, this scheme is only suitable for a limited number of large transactions but not for micro-transactions as it would become prohibitively expensive.

In order to tackle the limitations in blockchain 1.0 and 2.0, a third generation of blockchain platforms are currently under development such as Dfinity [10], NEO [8], IOTA [7] and Ethereum [15], using different approaches. They aim to support multiple programming languages and the development of various mobile based applications.

As the usage of blockchain is continuing to increase, a fourth generation of blockchain platforms is being proposed by Seele whose aim is to innovate the new era of Value Internet. Blockchain 4.0 (aka Seele [3]) proposed new consensus algorithms based on Neural Networks that improves the fault tolerance of the system. The proposal also includes a new network architecture, low latency internet connection protocol to enable integration with Internet resources and the development of blockchain-based services.

6   Blockchain-Based Internet and Its Challenges

In the previous section, it was observed in the different trends of blockchain development new systems are trying to address the problem of scalability and performance without sacrificing the security of the information maintained by the network. Keeping these trends in mind, we envision that blockchain technology will significantly advance and become the basis for building totally autonomous security systems that solve the privacy and trust issues faced in today's web era. However, for this realisation to occur there are several challenges that need to be addressed before the current blockchain technologies can simultaneously ensure scalability, privacy and reliability at scales with billions of transactions every second. Here, we note the key challenges:

- **Scalability:** The current blockchain requires all transactions to be stored and be available for validating any new transaction. Due to this, cryptocurrencies such as Bitcoin can only process a few transactions every second. Newer systems fail to scale after some threshold of record and network sizes. There are several sub-problems that must be solved to address the issues, for example the optimisation of storage for transactions requiring intelligent means to maintain only a minimal amount of data to validate transactions. This also involves the challenge of when data can be archived and deleted. In addition, how data should be distributed between different nodes to ensure the best efficiency and scalability is another challenge in this context.

When considering the scalability of blockchain networks, the consensus protocol also plays an important role. Therefore, load-balancing in terms of how many and which nodes should be used to validate every transaction among participating nodes is another important question to answer.

- **Interoperability of Multiple Blockchains:** Given the highly distributed and heterogeneous nature of the Internet, we can envisage there will be several private and public blockchains co-existing in the ecosystem. To maintain a global state of the information, these different blockchains should be able to communicate in a secure and transparent manner without affecting security. For example, to know the exact identity of a user, several blockchains may be queried before a blockchain validates the transaction of that user.

- **Blockchain and AI:** Current blockchain protocols are effective in securing and validating the information stored within the network, however most of these are simple and they require long verification times even though the number of nodes in the network is relatively small – the efficiency of current consensus protocols need to improve. However, at greater scales, we can expect millions of nodes and this increases the risk of malign nodes trying to break the system. Several AI algorithms can help solve this and many other problems by making different parts of the blockchain 'smarter'. For example, the behaviour of nodes can be learned through their different actions and communication interactions, enabling smart decision-making on whether a particular node is trust-worthy or not. Thus, if some nodes are not trustworthy, they can be automatically pruned from decision making and this reduces the cost of adding new block.

- **Energy:** The maintenance of a secure system also comes at a cost. In particular, it can be very energy intensive. The current Bitcoin ecosystem has been estimated to consume the electrical equivalent of some small cities. When we consider Internet-scale networks with heterogeneity of connection types and devices such as mobile phones, energy usage becomes a key factor. This requires intelligent management of data and computation depending on the device's capacity in terms of computation and battery power.

- **Simulation and Testing:** Recently several types of blockchain-based systems have appeared. Each claim to offer advantages over the others. There is currently no standardised simulation environment or benchmarks that are available than can allow comparison between different proposed concensus and data structures in addition to testing security concerns. Simulation environments are not only essential for testing the currently proposed systems but also for future development. Moreover, simulation environments are cost-effective and allow repeatability of results.

7   Conclusion

In this paper, we survey blockchain technologies and applications from different perspectives. We first include an overview of blockchain concepts, and then categorize and compare data structures used in blockchains. We also compare the consensus protocols, and summarize blockchain implementations. Furthermore, we outline some challenges that need to overcome to improve the privacy and security of current internet services. It is becoming clear from developments over the last few years, that blockchain applications have increased and with them come necessary modifications to blockchain's features to make it more scalable and fault tolerant. However, when using blockchain in massive-scale systems, scalability limitations become prevalent leading to poor performance when millions of nodes participate. Evolving design considerations will enable blockchain to be able to communicate and interoperate on networks of enormous size, maintaining a global and reliable repository of information. AI can address some of the problems that need to be solved, however its applicability needs to be tested in different scenarios. We will consider in depth the modification and ramifications of a combined blockchain and AI approach in future research.

References

1. Allcott, H., Gentzkow, M.: Social media and fake news in the 2016 election. J. Econ. Perspect. **31**(2), 211–236 (2017)
2. Anh, D.T.T., Zhang, M., Ooi, B.C., Chen, G.: Untangling blockchain: a data processing view of blockchain systems. IEEE Trans. Knowl. Data Eng. **30**, 1366–1385 (2018)
3. Bi, W.: Seele project (2018). https://seele.pro/
4. Buyya, R., Garg, S.K., Calheiros, R.N.: SLA-oriented resource provisioning for cloud computing: challenges, architecture, and solutions. In: 2011 International Conference on Cloud and Service Computing (CSC), pp. 1–10. IEEE (2011)
5. Castro, M., Liskov, B., et al.: Practical Byzantine fault tolerance. In: OSDI 1999, pp. 173–186 (1999)
6. Culnan, M.J., McHugh, P.J., Zubillaga, J.I.: How large us companies can use twitter and other social media to gain business value. MIS Q. Executive **9**(4) (2010)
7. Divya, M., Biradar, N.B.: IOTA-next generation block chain. Int. J. Eng. Comput. Sci. **7**(04), 23823–23826 (2018)
8. Eisses, J., Verspeek, L., Dawe, C., Dijkstra, S.: Effect network: decentralized network for artificial intelligence (2018)
9. Gramoli, V.: From blockchain consensus back to byzantine consensus. Future Gener. Comput. Syst. (2017)
10. Hanke, T., Movahedi, M., Williams, D.: DFINITY technology overview series consensus system (2018)
11. Kiayias, A., Russell, A., David, B., Oliynykov, R.: Ouroboros: a provably secure proof-of-stake blockchain protocol. In: Katz, J., Shacham, H. (eds.) CRYPTO 2017. LNCS, vol. 10401, pp. 357–388. Springer, Cham (2017). https://doi.org/10.1007/978-3-319-63688-7_12
12. Kwon, J.: Tendermint: consensus without mining (2014). Accessed 18 May 2017
13. McAfee, A., Brynjolfsson, E., Davenport, T.H., Patil, D., Barton, D.: Big data: the management revolution. Harvard Bus. Rev. **90**(10), 60–68 (2012)
14. Nakamoto, S.: Bitcoin: a peer-to-peer electronic cash system (2008)
15. Ethereum Project (2018). https://www.ethereum.org/
16. Ridyard, P.: Tempo white paper (2018). https://projects.radix.global/wiki/radix/wikis/Tempo-White-Paper

17. Schwartz, D.: The ripple protocol consensus algorithm (2018). https://ripple.com/files/rippleconsensuswhitepaper.pdf
18. Walport, M.: Distributed ledger technology: beyond blockchain. UK Government Office for Science (2016)
19. Zyskind, G., Nathan, O., et al.: Decentralizing privacy: using blockchain to protect personal data. In: 2015 IEEE Security and Privacy Workshops (SPW), pp. 180–184. IEEE (2015)

Selective Comprehension for Referring Expression by Prebuilt Entity Dictionary with Modular Networks

Enjie Cui[1], Jianming Wang[1,2], Jiayu Liang[2], and Guanghao Jin[2(✉)]

[1] School of Electronic and Information Engineering,
Tianjin Polytechnic University, Tianjin, China
`cuienjiecv@163.com`
[2] School of Computer Science and Software Engineering,
Tianjin Polytechnic University, Tianjin, China
`yyliang2012@hotmail.com`, {`wangjianming,jinguanghao`}`@tjpu.edu.cn`

Abstract. Referring expression comprehension, known as the technique of localizing entities in an image based on natural language expression, is still a challenging task far from solved. In literature, researchers always focused on how to localize the correct image region according to a natural language expression and never questioned the correctness of the expression. In practical scenarios, the situation is common. For example, there is a pumpkin on the table, but the expression is "there is a watermelon on the table". It is obvious that incorrect location can be derived from a wrong expression, which state-of-the-art approaches cannot avoid. In this paper, we propose modular networks to solve this problem, which includes three main parts, i.e. the expression filtering module, the expression analysis module and the localization module. Specifically, the expression filtering module adopts an entity dictionary to list all the objects in the image, which is prebuilt by an object detection method, to discriminate whether an expression is correct or not. In this way, our model realizes selective comprehension of referring expression, which can output a "wrong expression" feedback instead of a wrong image region localization when an expression is determined as wrong. Sufficient experiments shows that our model can efficiently filter wrong expressions and effectively solve the problem of referring expression compression in practical scenarios.

Keywords: Referring expression · Selective comprehension
Entity dictionary · Modular networks

This work was supported by National Natural Science Foundation of China (Grant No.61373104) and Natural Science Foundation of Tianjin (Grant No.16JCYBJC42300 and Grant No.17JCQNJC00100) and Science and Technology Commission of Tianjin Municipality (Grant Nos.15JCYBJC16100) and Program for Innovative Research Team in University of Tianjin (No. TD13-5032).

K. Yoshida and M. Lee (Eds.): PKAW 2018, LNAI 11016, pp. 211–220, 2018.
https://doi.org/10.1007/978-3-319-97289-3_16

1   Introduction

Referring expression is a special case of natural language expression, which uses a simple and effective description to express entities in the images [7]. Referring expression is used to describe the specific entities in the physical environment according to the unique attributes. For example, "The apple on the table" contains the two attributes of "apple" and "table". Referring expression comprehension becomes a hot topic in both computer vision and natural language processing in these years, which objective is to locate the described entities in the images by the information of the referring expression [4,9,10,12].

At present, most of the existing models use the features of the candidate region to obtain the probability of matching the referring expression, and then output the candidate region with the highest probability as the predicted result [3,7,8,14]. However, in realistic scenarios, the expression is not always accurate or reliable. For example, the user commands the robot to go to other scenes to pick something. There may be errors in the user's memory or the scene may have changed for some reason. In this case, the user's expression is very likely to be wrong. In addition, misunderstandings of objects by users can also lead to incorrect expression. Therefore, it is very common for questioners to have wrong expressions in practical scenarios.

The existing methods focus on how to improve the performance of referring expression comprehension, but doesn't consider the correctness of the expression. As we can see in Fig. 1, there is a pumpkin on the table. The correct expression should be like "The pumpkin on the table". For some reason, it might be "The watermelon on the table." In this case, the existing module would get the wrong location based on the wrong expression, thus misleading the questioner. Therefore, it is very important to judge the correctness of the expression in practical scenarios.

In this paper, we propose a modular networks to solve the problem. Our network consists of three modules: expression filtering module, expression analysis module and localization module as we can see in Fig. 2. The difference between existing methods and our method is that we designed an expression filtering module to check the correctness of the expression. Our expression filtering module uses R-FCN [2] to get the information of the objects in the image and prebuilt an entity dictionary. Then uses the entity dictionary to judge whether referring expression correctly describes the objects in the image. If the model judges that the expression is correct, perform the locating task. If it is judged that the expression is wrong, the information is fed back to the user and the user is required to re-express. Our modular network can effectively determine whether the referring expressions are "correct" or "wrong" and only perform the locating task in "correct" case. Therefore, our model can achieve selective comprehension of referring expression.

The major contribution of our work consist is that we proposed a referring expression comprehension problem in practical scenarios that may have wrong expression. We use a modular networks to solve the problem which has the expression filtering module. We evaluate our model on multiple datasets, and the

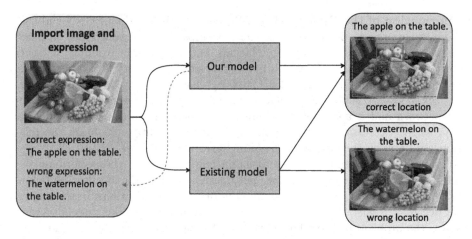

Fig. 1. The left box indicate the referring expression comprehension in practical scenarios where the expression of the input has wrong expression. The box on the right indicates that the model obtains correct and wrong location results based on different correctness expressions. Existing methods will get wrong location based on wrong expression and mislead the questioner. Our model can solve this problem and feedback information for "wrong expression" was shown in dotted lines.

experiments show that our model can successfully solve the problem of referring expression comprehension in practical scenarios and achieves better performance than the existing methods.

2   Related Work

Referring Expression Comprehension. The problem of referring expression comprehension can be summarized as a retrieval problem over image regions [3,4,9,10,12,15]. First, a series of candidate regions of the objects in the image are extracted by the object proposal methods [5,13,17]. Then, the matching degree between candidate area and referring expression is scored and a candidate area that has the highest score is returned as the localization result. In paper [4,9], the score of each candidate area is based on the local visual features of the candidate region and the context feature of the whole image. The other methods [3,10,15] doesn't only extract the local visual features of a single area, but also consider multiple candidate areas. Adding contextual features extracted from other candidate regions in the image, those methods can locate a referring expression into a pair of regions.

Generally, recurrent neural network is used to process the natural language which can be divided into three kinds: the first one is through the distribution of referring expression to predict [3,4,8,9]; second one is the referring expression encoding into a vector representation (vector representation [10,12]); The third one is to use attention model to parse the expression into three-tuple form (subject, relation, object), and then use modular network to solve that task.

Information of the Relationships. Recent work is based on RCNN and uses a linguistic prior to detecting visual relationships [1]. This work relies on fixed, predefined categories for subjects, relations, and objects. For example, it treats entities like "desk" as subjects or objects and preposition "on" as discrete classes. Compositional Modular Networks (CMNs) [3] has been proved efficient in referring expression comprehension. Generally, the network is designed as a modular network and it also considers multiple candidate regions, uses the analytic expression (expression parsing) and the local visual entity to obtain the relationship between the objects that are included in expression.

Most of the existing methods focus on how to improve the performance of referring expression comprehension, but doesn't consider the localization failure by wrong expression, which are common in practical applications. In this paper, we design a robust reference expression comprehension model that effectively solves the problem by modular networks with expression filtering module.

3   Our Model

Our modular networks consist of three sub-modules: the expression filtering module, the express parsing module, and the localization module. The three sub-modules solve the different problems respectively. The core of our network is expression filtering module, which can determine whether the expression is correct or not, and transfer the correct expression to the express parsing module. The express parsing module can learn to parse the expressions into these a 3-component triplet (subject, relationship, object) with attention and extract three vector representations corresponding to these three components. The localization module using local visual features and global contextual features of the image, and consider multiple regions and output the localization result which has highest probability. Figure 2 shows an overview of our model.

3.1   Expression Filtering Model

The expression filtering module uses the retrieval method to complete the filtering task. First, we train a R-FCN model for object detection on the MSCOCO dataset [6]. R-FCN is a region-based object detection framework leveraging deep fully-convolutional networks, which is accurate and efficient. We use R-FCN to obtain the information of the objects in the image and create an entity dictionary. Then we parse referring expression into independent words and compare those with entity dictionary. If there are two or more words that can be found in the entity dictionary, it means that the probability of correct expression is high and our model determined that the expression is correct, otherwise the expression is wrong. The number of matching numbers is set to two because there are at least two object attributes of subject and object in the correct expression that corresponds to the elements of the entity dictionary as we show in Fig. 3.

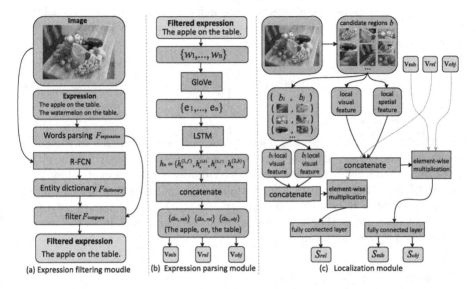

Fig. 2. (a) The expression filtering module can determine whether the expression is correct. (b) The expression parsing module can parse the correct expression into (subject, relationship, object) and encoding them into corresponding vectors. (c) The localization module locates the object in the image.

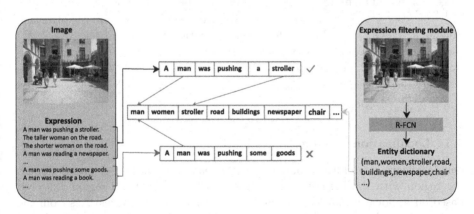

Fig. 3. Illustration of expression filtering module. The left box indicate the input image and expressions, where there are correct and wrong expressions. The right box indicate that the corresponding entity dictionary is created based on the image. The expression filtering module determines whether the expression is correct based on the entity dictionary.

We create an entity dictionary $F_{dictionary}(J) = \{d_1 \ldots d_j\}$ based on the input image I. Then we parse the expression into independent words $F_{expression}(I) = \{w_1 \ldots w_i\}$. We define the Comparison function of word w_i with dictionary element d_j as below:

$$F_{compare}(w_i, d_j) = \begin{cases} 1, & w_i = d_j \\ 0, & w_i \neq d_j \end{cases} \tag{1}$$

If $\sum_{i=1}^{n} \sum_{j=1}^{m} F_{compare}(w_i, d_j) < 2$, it proves that at least one of the subject and the object in the expression doesn't conform to the elements in the entity dictionary. The probability that such an expression correctly describes the object in the image is relatively low, and model determine it is a wrong expression. Then our model gives users feedback information, which indicates that the expression is wrong and filter it out. On the contrary, if $\sum_{i=1}^{n} \sum_{j=1}^{m} F_{compare}(w_i, d_j) \geq 2$, our model determines the expression to be correct and pass it to the module behind to complete the locate task.

3.2 Expression Parsing Module

The express parsing module can decompose the expression E into the a triplet (subject, relationship, object), and generate related vector representations v_{sub}, v_{rel}, v_{obj} from E through a soft attention mechanism over the word sequence as we can see in Fig. 2(b).

It first converts E into an independent word $\{w_n\}_{n=1}^{N}$, and then embed each word w_n to a vector e_n using GloVe [11]. Next it uses a two-layer LSTM network to get all the hidden states and concatenated them into a single vector $h_n = [h_n^{(1,f)} h_n^{(1,b)} h_n^{(2,f)} h_n^{(2,b)}]$. Then it uses the connection layer to obtain the attention weights $a_{n,sub}$, $a_{n,rel}$ and $a_{n,obj}$ for subject, relationship and object.

Finally three component triplet representations of the subject v_{sub}, relationship v_{rel}, and object v_{obj} are extracted as word embedding vectors e_n with attention weights as $v_{sub} = \sum_{n=i}^{N} a_{n,sub} e_n$, $v_{rel} = \sum_{n=i}^{N} a_{n,rel} e_n$ and $v_{obj} = \sum_{n=i}^{N} a_{n,obj} e_n$.

3.3 Localization Module

Our localization module outputs a score for each candidate region and returns the region with the highest score as localization result. Each region is scored based on its local visual features and some global contextual features from the whole image. The score of a region presents whether it matches the expression of subject or object. It considers multiple regions at the same time as we illustrate in Fig. 2(c). Then relationship between those regions is also checked.

Given an image, we use an object detector to locate a set of candidate regions b. Each candidate region comes with a bounding boxes. The next step is to produce a set of region pair (b_i, b_j) from the detected objects. With n detected objects, we can form $n(n-1)$ pairs. Our localization module solves two different sub-tasks: single region model $f_{sig}(b_i, v_{sub}; \Theta_{sub})$ and $f_{sig}(b_j, v_{rel}; \Theta_{rel})$ for deciding whether a single region matches the subject or object in the expression, where v_{sub} and v_{obj} is the textual vector representation of the subject and object. Another is pair of regions model $f_{pair}(b_i, b_j, v_{rel}; \Theta_{rel})$ for deciding whether a pair of regions matches the relationship described in the expression represented by v_{rel}.

We define the score over a pair of image regions matching an input referential expression as the sum of three components:

$$S(b_i, b_j) = f_{sig}(b_i, v_{sub}; \Theta_{sub}) + f_{sig}(b_j, v_{obj}; \Theta_{obj}) + f_{pair}(b_i, b_j, v_{rel}; \Theta_{rel}) \quad (2)$$

Finally, we take the highest score (b_i, b_j) as output.

$S_{sub} = f_{sig}(b_i, v_{sub}; \Theta_{sub})$ and $S_{obj} = f_{sig}(b_j, v_{obj}; \Theta_{obj})$ representing how likely a region bounding box b_i and b_j matches v_{sub} and v_{obj}. This module takes the local visual feature x_{vis} and spatial feature $x_{spatial}$ of image region b. Then x_{vis} and $x_{spatial}$ are concatenated into a vector $x_{v,s} = [x_{vis}x_{spatial}]$ as representation of region b. Since element-wise multiplication is shown to be a powerful way to combine representations from different modalities [1], we adopt it here to obtain a joint vision and language representation.

In our implementation, $x_{v,s}$ is embedded to a new vector $\tilde{x}_{v,s}$ using $W_{v,s}$ and $b_{v,s}$. Then element-wise multiplied with v_{sub} and v_{obj} to obtain a vector z_{sub} and z_{obj}. Finally the score s_{sub} and s_{obj} is predicted linearly from z_{sub} and z_{obj} using w_{sub}, b_{sub}, w_{obj} and b_{obj}. So that the parameters in Θ_{sub} and Θ_{obj} are $(W_{v,s}, b_{v,s}, w_{sub}, b_{sub})$ and $(W_{v,s}, b_{v,s}, w_{obj}, b_{obj})$.

$S_{rel} = f_{pair}(b_i, bj, v_{sub}; \Theta_{sub})$ representing how likely a pair of region bounding boxes (b_i, b_j) matches v_{rel}, the representation of relationship in the expression. we use the spatial features x_1 and x_2 of the two regions b_1 and b_2 extracted in the same way as in single region module. Then x_1 and x_2 are concatenated as $x_1, x_2 = [x_1 x_2]$, and then processed in a similar way as in single region module to obtain s_{rel}, The parameters in Θ_{rel} are $(W_{1,2}, b_{1,2}, w_{rel}, b_{rel})$.

4   Experiments

We apply our method to images and expressions of the Google-Ref dataset [9] which is a benchmark dataset for referring expression comprehension. We also evaluate our model on the Visual-7W dataset [16].

4.1   Add Error Expression

We set up test dataset of the wrong-expression by following steps. We use some predesigned wrong expressions to replace the half of the expressions in the Google-Ref and Visual-7W dateset respectively. Thus the probability of wrong and correct expression is same, so that, there is no bias between two kinds expressions.

4.2   The Evaluation on Google-Ref Dataset

We first evaluate our model on the Google-Ref dataset, which contains 104560 expressions referring to 54822 objects from 26711 images selected from MS COCO. As the dataset does not explicitly contain subject-object pairs annotation for the referring expressions, we train our model with weak supervision by optimizing the subject score s_{sub} using the expression-level region ground-truth. The candidate bounding box set B at both training and test time are all the annotated entities in the image.

Table 1. Expression filtering module (EFM)'s accuracy about filtering in Google-Ref dataset and Visual-7W dataset (both add wrong-expression in test dataset).

Dataset	EFM accuracy
Google-Ref	97.6%
Visual-7W	86.7%

Table 2. Top-1 precision of our model and existing models on Google-Ref dataset (add wrong-expression in test dataset).

Method	Mao [9]	Yu [15]	Nagaraja [6]	CMNs [3]	Our model
P@1	41.5%	43.7%	46.9%	47.6%	61.9%

We first tested the accuracy of the expression filtering module on Google-Ref dataset. The experimental result is shown in Table 1. The expression filtering module performs well on Google-Ref dataset which accuracy is 97.6%.

In the test phase, we evaluate on the dataset using the top-1 precision (P@1) metric, which is the fraction of the highest scoring subject region matching the ground-truth for the expression. Table 2 shows the performance of our model and the existing models. We can conclude that our model has obvious better performance.

4.3   The Evaluation on Visual-7W

Then, we evaluate our method on Visual-7W dataset which has the multiple choice pointing questions in visual question answering. Given an image and a question like "Which barn door is behind the man", the task is to select the corresponding region from a few choice regions as an answer. Since this task is closely related to referential expressions. Our model can be trained in the same way as in Google-Ref dataset to score each choice region using subject score and pick the highest scoring choice as answer.

We pre-trained our model with weak supervision and extract visual feature by MSCOCO-pretrained Faster-RCNN VGG-16 network extract visual feature.

Table 3. Accuracy of our model and the existing models in Visual-7W dataset (add wrong-expression in test dataset).

Method	Zhu [16]	CMNs [3]	Our model
Accuracy	30.4%	39.3%	72.5%

Here we use two different candidates bounding box sets B_{sub} and B_{obj} of the subject regions and the object regions, where B_{sub} is the 4 choice bounding boxes, and B_{obj} is the set of 300 proposal bounding boxes extracted using RPN in Faster-RCNN [18].

The test phase is the same as that of the Google-ref dataset. As shown in Table 1. We also tested expression filtering module's accuracy in Visual-7W. The accuracy is 86.7%. Table 3 shows that our model achieves higher accuracy than the existing methods. We can draw a conclusion that our model has a very good performance in handling referring expression comprehension which may contain wrong expressions for practical scenarios.

5 Conclusion

In this paper, a new referring expression comprehension model was proposed in practical scenarios, which solves the wrong expression problem that hasn't been considered by the existing methods. If a referring expression doesn't correctly describe the image, it may lead to a wrong location, which can mislead questioners. Our model can conduct selective comprehension of referring expression based on a prebuilt entity dictionary that is generated by an object detection method. Sufficient experiments show that our method can efficiently filter the wrong expression and effectively solve the problem of referring expression compression in practical scenarios. In the future, the relationship between object pairs in images will be studied to establish a relational dictionary, which can judge the correctness of the expression more effectively.

References

1. Ba, J., Mnih, V., Kavukcuoglu, K.: Multiple object recognition with visual attention. arXiv preprint arXiv:1412.7755 (2014)
2. Dai, J., Li, Y., He, K., Sun, J.: R-FCN: object detection via region-based fully convolutional networks. In: Advances in Neural Information Processing Systems, pp. 379–387 (2016)
3. Hu, R., Rohrbach, M., Andreas, J., Darrell, T., Saenko, K.: Modeling relationships in referential expressions with compositional modular networks. In: 2017 IEEE Conference on Computer Vision and Pattern Recognition (CVPR), pp. 4418–4427, July 2017. https://doi.org/10.1109/CVPR.2017.470
4. Hu, R., Xu, H., Rohrbach, M., Feng, J., Saenko, K., Darrell, T.: Natural language object retrieval. In: 2016 IEEE Conference on Computer Vision and Pattern Recognition (CVPR), pp. 4555–4564, June 2016. https://doi.org/10.1109/CVPR.2016.493

5. Krähenbühl, P., Koltun, V.: Geodesic object proposals. In: Fleet, D., Pajdla, T., Schiele, B., Tuytelaars, T. (eds.) ECCV 2014. LNCS, vol. 8693, pp. 725–739. Springer, Cham (2014). https://doi.org/10.1007/978-3-319-10602-1_47

6. Lin, T.-Y., et al.: Microsoft COCO: common objects in context. In: Fleet, D., Pajdla, T., Schiele, B., Tuytelaars, T. (eds.) ECCV 2014. LNCS, vol. 8693, pp. 740–755. Springer, Cham (2014). https://doi.org/10.1007/978-3-319-10602-1_48

7. Liu, J., Wang, L., Yang, M.H.: Referring expression generation and comprehension via attributes. In: 2017 IEEE International Conference on Computer Vision (ICCV), pp. 4866–4874, October 2017. https://doi.org/10.1109/ICCV.2017.520

8. Luo, R., Shakhnarovich, G.: Comprehension-guided referring expressions. In: 2017 IEEE Conference on Computer Vision and Pattern Recognition (CVPR), pp. 3125–3134, July 2017. https://doi.org/10.1109/CVPR.2017.333

9. Mao, J., Huang, J., Toshev, A., Camburu, O., Yuille, A., Murphy, K.: Generation and comprehension of unambiguous object descriptions. In: 2016 IEEE Conference on Computer Vision and Pattern Recognition (CVPR), pp. 11–20, June 2016. https://doi.org/10.1109/CVPR.2016.9

10. Nagaraja, V.K., Morariu, V.I., Davis, L.S.: Modeling context between objects for referring expression understanding. In: Leibe, B., Matas, J., Sebe, N., Welling, M. (eds.) ECCV 2016. LNCS, vol. 9908, pp. 792–807. Springer, Cham (2016). https://doi.org/10.1007/978-3-319-46493-0_48

11. Pennington, J., Socher, R., Manning, C.: Glove: global vectors for word representation. In: Proceedings of the 2014 Conference on Empirical Methods in Natural Language Processing (EMNLP), pp. 1532–1543 (2014)

12. Rohrbach, A., Rohrbach, M., Hu, R., Darrell, T., Schiele, B.: Grounding of textual phrases in images by reconstruction. In: Leibe, B., Matas, J., Sebe, N., Welling, M. (eds.) ECCV 2016. LNCS, vol. 9905, pp. 817–834. Springer, Cham (2016). https://doi.org/10.1007/978-3-319-46448-0_49

13. Uijlings, J.R., Van De Sande, K.E., Gevers, T., Smeulders, A.W.: Selective search for object recognition. Int. J. Comput. Vis. 104(2), 154–171 (2013)

14. Yu, L., Tan, H., Bansal, M., Berg, T.L.: A joint speaker-listener-reinforcer model for referring expressions. In: 2017 IEEE Conference on Computer Vision and Pattern Recognition (CVPR), pp. 3521–3529, July 2017. https://doi.org/10.1109/CVPR.2017.375

15. Yu, L., Poirson, P., Yang, S., Berg, A.C., Berg, T.L.: Modeling context in referring expressions. In: Leibe, B., Matas, J., Sebe, N., Welling, M. (eds.) ECCV 2016. LNCS, vol. 9906, pp. 69–85. Springer, Cham (2016). https://doi.org/10.1007/978-3-319-46475-6_5

16. Zhu, Y., Groth, O., Bernstein, M., Fei-Fei, L.: Visual7W: grounded question answering in images. In: 2016 IEEE Conference on Computer Vision and Pattern Recognition (CVPR), pp. 4995–5004, June 2016. https://doi.org/10.1109/CVPR.2016.540

17. Zitnick, C.L., Dollár, P.: Edge boxes: locating object proposals from edges. In: Fleet, D., Pajdla, T., Schiele, B., Tuytelaars, T. (eds.) ECCV 2014. LNCS, vol. 8693, pp. 391–405. Springer, Cham (2014). https://doi.org/10.1007/978-3-319-10602-1_26

18. Ren, S., He, K., Girshick, R., Sun, J.: Faster R-CNN: towards real-time object detection with region proposal networks. IEEE Trans. Pattern Anal. Mach. Intell. 39(6), 1137–1149 (2017)

Pose Specification Based Online Person Identification

Tao Guo[1], Jianming Wang[1,2], Rize Jin[2], and Guanghao Jin[2(✉)]

[1] School of Electronic and Information Engineering, Tianjin Polytechnic University,
Tianjin, China
1727336459@qq.com
[2] School of Computer Science and Software Engineering, Tianjin Polytechnic
University, Tianjin, China
{wangjianming,jinguanghao}@tjpu.edu.cn, rizejin@kaist.ac.kr

Abstract. Generally, identification methods use high quality frames that have obvious features like whole face of human being. In human identification case, multiple recognition areas have been proved to be a significant improvement over traditional face recognition methods. The main challenge of human recognition are that in some poses, the identification leads to a result of low accuracy as there are no obvious features like a whole face. In order to solve that problem, we apply the networks to detect the additional information to process the images that are hard to be used for identification. In continuous online conditions, there may still be some frames that can not be detected with those efforts. Our method uses a weight system to record changes in posture. Then the sequence of frames that belong to the same person can be grouped and the undetected frames can be identified by the detected frames. Experimental results show that our model achieves higher recognition accuracy than the existing methods in online case.

Keywords: PSM · Person indentification · Online

1 Introduction

The task of human face recognition, fingerprint recognition, speech recognition, etc. have been successfully solved with the deep learning networks. The detection accuracy of crime detection and security inspection systems is a important applications which uses those technologies. Compare with other biometric identification technologies, face recognition are The practicality has unique technical

This work was supported by National Natural Science Foundation of China (Grant No. 61373104) and Natural Science Foundation of Tianjin (Grant No. 16JCY-BJC42300 and Grant No. 17JCQNJC00100) and Science and Technology Commission of Tianjin Municipality (Grant Nos. 15JCYBJC16100) and Program for Innovative Research Team in University of Tianjin (No. TD13-5032).

K. Yoshida and M. Lee (Eds.): PKAW 2018, LNAI 11016, pp. 221–230, 2018.
https://doi.org/10.1007/978-3-319-97289-3_17

advantages. Therefore, it is highly concealed and is suitable for security prevention, criminal monitoring, and criminal pursuit. The biggest advantage of face recognition is that it is reliable than other information.

However, the face is not easily captured for some reason like the position of the camera or some obstacles. The recognition [1]. In many actual scenes, the faces are often invisible, and recognition of people in those scenes may produce erroneous results, resulting in low recognition accuracy. In addition, in movies or sports videos, people's faces may also be obstructed. The low resolution of the frames also causes difficulty in recognition. People may move away from the camera or turn in different directions as shown in the Figs. 1 and 2. Therefore, in order to obtain more identification cues, we need more regional information for identification. Recent studies have shown that different parts of the body can provide complementary information for identification and can significantly improve the accuracy of human recognition.

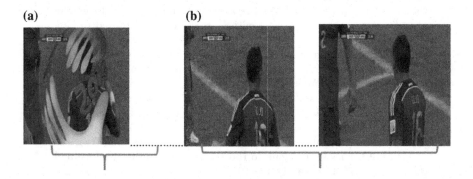

Fig. 1. (a) cannot be identified (b) can be identified by character and color

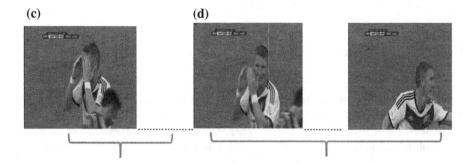

Fig. 2. (c) cannot be identified (d) can be identified by face

In online case, there are still some frames that can not be detected with different identification networks. To solve that problem, we divide the human body into different regions, and each region contains samples of specific human orientation. Then we use a multi-region convolutional neural network to represent each pose, and a weight system is to show the possibility of poses. Since we can predefine the 7 most basic poses, the possibility of poses can show the pose details that are not included in the pre-defined poses. At the same time, we also added the LSTM network to process the pictures in consecutive frames and filter out the sequence that can be used for recognition so as to improve the efficiency and accuracy in the recognition process [2].

In order to solve the problem of low recognition accuracy, we mainly do the following efforts: First, the images in the data set are classified in different poses, and different poses are extracted for each frame. The key feature are that different poses are trained separately to obtain different PSMs. Then we use the LSTM networks to process pictures to obtain pictures that can be grouped for the same person. Pose perception weights are based on different PSM multiple classifiers combined by the pose perception weights provided by the pose estimator, are finally identified through the recognition system. Then we use the networks to identify a person with the information of face, character and color of clothes. The additional information can help our method identify a person. As the frames are grouped for the persons, the frames that can not be identified by the networks can be covered by the identified frames when those frames belong to the same person. The experiment result shows that our method can achieve higher accuracy than the existing methods.

2   Related Work

The main task of **person re-identification** are to perform pedestrian matching in non-overlapping camera views. The main application are in video surveillance systems. Until now, the most popular functions have been evaluated and measured by manual or data-driven. Used to achieve visual, pose, and photo metric conversion differences. These methods also optimize the joint architecture for re-identifying consecutive losses in non-overlapping body regions.

People recognition with multiple body cues is interested in the work. We directly compared the recent efforts to use a variety of body cues: Li PIPER, Naeil, and PIPER USES a complex pipeline containing 109 classifiers, each of which is based on different body part [3]. Including Deep image training based on millions of Face structure of a kind of representation, a systemic training Alex Net and 107 posture training - patch Alex Nets, two after using PIPA training set. On the other hand, naeil is based on a fixed body area such as the face, head and body and four different sets of data set training scenarios and character cues using PIPA and PETA [4]. Although the pose-let (for PIPER) normalizes the posture, they have a poor sense of discrimination compared to the fixed body areas used by naeil. We combine the advantages of these two methods based on the postural perception of the fixed body region.

We focus on the generic person recognition problem similar to that work in diverse settings without using any domain level information and demonstrate the effectiveness of the pose-aware models in different scenarios.

3   Our Method

The framework of our proposed approach is shown in Fig. 3. The biggest challenge in the process of character recognition is that the subject can not be identified in some pose. The physical appearance of the recognition targets also different with the change of the posture. We solve this problem mainly by learning the posture-specific model (PSM). PSM corresponds to a specific discriminative features related to a particular gesture [5]. We first pass a continuous picture through the LSTM network, group the pictures that belong to same person [6]. Then, we use two CNN networks for the player face, number and color of cloths respectively. By the grouped frames for the same person, we can cover the parts that cannot be used for identification by the detected frames, thereby we can improve the recognition accuracy [7].

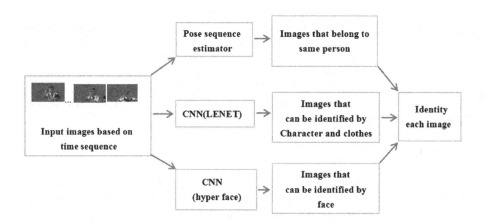

Fig. 3. Our approach: Input continuous images, through the pose estimator, output images that belong to the same person. Through the LENET network, our method output images that can identify the shirt number, through the hyper face network, our method output pictures that can be used to identify the face. Then a person can be identified and remaining images can be covered by this information.

3.1   Pose Prediction and Recognition System

The framework of our proposed framework is shown in Fig. 4. We cluster those images into a set of posture vies based on the assignment of the labels. Learn pose estimators on these clusters for view classification. To learn the characterization of a person in each view, the PSM is then trained for identification using multiple body regions [8]. We trained multiple linear classifiers based on PSM representations for predicting identities. Similar to paer, for a given an input image x, we first compute pose-specific identity scores $S_i(y,x)$, each based on the i-th PSM representation. The final score for each identity y are a linear combination of the pose-specific scores, and the formula for calculation is as follows: [9]

$$S(x,y) = \sum_i WiSi(y,x) \tag{1}$$

In view of different body parts, we train individual posture points in each of these areas, but although distinguishing the body areas can contribute to the improvement of the recognition accuracy, there may be differences due to the training scene. For example, when we detect that the player can identify, his body is in a half-left state, and at this time, he contains a less-informed area of the shielding surface. At the same time, in other cases, when we can only capture the player's face, he can provide less upper body information for recognition at this time. We train linear SVM classifiers for each of these eigenvectors to obtain identity predictions. The identity scores result $Si(y,x)$ for a particular pose is simply the sum of the outputs of the three SVM classifiers [10].

$$Si(y,x) = \sum_f Pi(y \mid f;x) \tag{2}$$

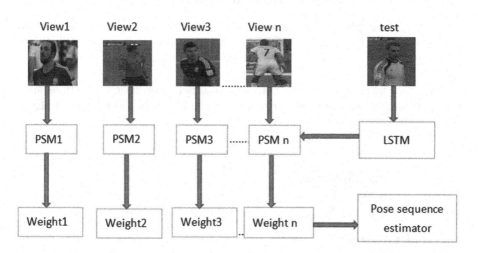

Fig. 4. Firstly classify images in datasets into different poses, extract key features of different poses, and train different poses to obtain different PSMs. We use the LSTM network to group the frames that belong to same person.

3.2   PSM and LSTM

We propose an approach to improve the generalization ability by allowing the network to selectively focus on informative body regions during the training process. The idea is to optimize both the head and upper body networks jointly over a single loss function. Our PSM contains two AlexNets corresponding to the head and upper-body regions. The final fc7 layers of each region are concatenated and passed to a joint hidden layer (fc7 plus) with 2000 nodes before the classification layer [11]. This provides more flexibility to the network to make the predictions based on one region even if the other region is noisy or less informative.

We use the LSTM network to process continuous pictures during the test. When entering a continuous picture, we use the LSTM network. Because the LSTM network has a short-term memory function, we use the LSTM network to group frames belonging to the same person. This aspect can improve the efficiency in the testing process. In addition, we can also improve the accuracy of recognition. The main advantage of improving accuracy are to use the short-term memory function of the LSTM networks to filter out the pictures that can be used for identification because we are Through different gestures to identify consecutive pictures, we mainly train in the seven most basic poses, but we have 7 gestures other than the basic gestures during the test, which makes us unable to identify them. The accuracy of recognition is not high, so we can use the LSTM networks to screen out the poses that can be used for recognition, which can improve the accuracy in the recognition process.

3.3 Face, Character and Clothes Recognition

When we perform player detection, we first use a network hyper face network to identify a person when there is clear face which is similar to paper [12]. When there are no face in the frames, we use LENET network to detect the number and clothes.

The main framework for the detection of other information situations such as player number, age, and face are shown in the Fig. 5, In the process of player

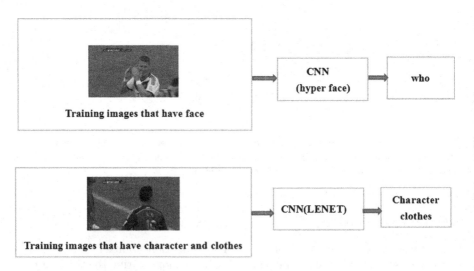

Fig. 5. We use the hyper face network to identify a person when the image contains a face. When there is no face, we enter the image that contains the player number into the LENET network to identify a person.

number detection, we mainly use the CNN network model by entering a continuous segment containing the player's number. In the picture, the player's number is output after processing by LENET. We use the output information to identify the player. In addition, we use other situations to process unrecognizable pictures, for example, we use a hyperface CNN convolutional network to input a continuous picture containing facial expressions. After processing by the hyperface convolution network, we output the result. The player's age, facial features and a series of information conditions, we identify the player by processing the information [13]. By combining the processing in the above two cases, we can further identify the unidentifiable player. For the pictures that are still can not identify through those networks, we use the group information of those frames, then the undetected frames can be covered by the result of detected frames.

4   Experiment

4.1   Soccer Dataset

We created a football data set from the video game between Argentina and Germany in the 2014 FIFA World Cup finals. We only considered the video clips in the video game. There are 37 video clips in our football data set, and the average duration of each video clip are 30 s. It includes 13 players from Germany, 14 players from Argentina and one referee [14]. For all our experiments, we have trained each PSM separately on the front, half left, right, left, right, back and partial views. Each PSM corresponds to a specific feature of a particular posture. The other seven models are trained on the other models, combined with the body's posture, and combined by the posture perception weights provided by the posture estimator [15]. Finally, the identification system is combined with the posture prediction system to complete the player's recognition process in order to recognize us. The accuracy was optimized and improved. We added LSTM network to improve and optimize during the test. We focused on a continuous frames by using the LSTM network. Since the LSTM network has a short-term memory function, we can filter out the poses that belong to different person. The continuous pose frames that belong to the same person changes in the same rules which are significantly different from those belong to the other person [16]. The LSTM can achieve the objective of detecting the poses that belong to different person. We train the network for a total of 300,000 iterations. For the training SVM and the basic weight Wi set from 0 to 1, parameter C is set to 1 [17].

4.2   Results and Analysis

Due to the limitations of the environment in the soccer data set, the resolution in the video clip is low, the motion environment is blurred, serious external disturbances and characters change continuously, and the accuracy of the character recognition is not high. We also observed in the course of the experiment that in the football data set, due to unusual postures (kicking, wrestling, barbs, etc.), it was impossible to train unusually on the basis of the 7 basic models during the training process. In addition, we used the combination of PSM and LSTM models in the original basic model, first classify the images in the dataset in different poses, extract the key features of different poses, and train different poses to obtain different PSMs. We use a test set to measure the performance of a gesture-specific PSM model. We use PSM and basic models to extract the features of these examples. Table 1 shows the performance of each model when different views are identified. It shows that for the forward, left, half, and right half examples, the corresponding PSM model is superior to other models, and the test results are shown in the following table:

Table 1. The performance of test in a particular pose represented using the different PSMs.

Pose	Model 0 (Right)	Model 1 (Semi-right)	Model 2 (Frontal)	Model 2 (Semi-left)	Model 3 (left)
Pose0 (right)	58.6	54.2	52.6	51	45.3
Pose1 (semiright)	48.4	65.6	64	64	49
Pose2 (frontal)	59.6	76.8	79.1	75.2	62
Pose3 (semileft)	54.6	69	68.4	72	58.4
Pose4 (left)	52.3	59	62	57.4	53.6

We test the impact on recognition results by adding tracking tags to the football dataset. We redistribute the frame tags in all frames in a dataset folder. The test results show that useful tracking information can be used to improve the recognition process. During the test, we introduced the Short Term Memory Network (LSTM). Due to the temporal correlation in the video, we use the short-term memory network (LSTM) to screen out the pictures that can be used for identification, improving the efficiency and accuracy of the test process. We are also in the front, left, left, and right The right half, back and partial views train the PSM separately. In addition, we used two networks to process consecutive pictures that could not be recognized in the method recognition by combining gestures. We used the hyper face network to output the pictures containing faces, which players' information we output. Enter the picture containing the player number into the LENET network, output the player's number, we use the number, player information and other relevant information to identify unrecognized continuous pictures, we pass on the number and player age, etc. The acquisition of information can make the recognition accuracy reach 79.68%. Also, we use less time and efficiency in testing, and the results of our tests are compared with other methods Shown (Table 2):

Table 2. Performance comparison on Soccer dataset

Method	Accuracy Without tracks	Accuracy With tracks
Head (H)	18.04	20.16
Upper body(U)	17.89	19.54
Separate training H and U	17.54	20.36
Joint training of H and U	18.46	20.54
naeil	19.48	23.67
PSM	21.23	24.79
Our approach	35.64	36.68

5   Conclusion

In this article, we have established a model for identifying the human body in different postures. Our model classifies poses into a limited category and uses a weight system to identify the frames that belong to same person. Then we use two networks to identify a person by face and additional information. In online applications, continuous frame recognition is more reliable because video quality cannot be guaranteed. Since the accuracy depends on the person's posture, our model uses the LSTM network to locate the sequence of frames and group them for each person. In addition, we also use the LENET network and the hyper-face network to detect other information such as the number and player's face, age, etc., respectively, to improve the accuracy in the recognition process. Through those efforts, we have established a model that can identify a person in online case which may have low quality for identification. In other words, when we cannot identify a person, we can use the information of the time sequence and use the indentified frames, so that, any frames can be detected. Our work can be used as the basis for online recognition for human body recognition in online situations.

References

1. Ge, L., Liang, H., Yuan, J., Thalmann, D.: 3D convolutional neural networks for efficient and robust hand pose estimation from single depth images. In: IEEE Conference on Computer Vision and Pattern Recognition, pp. 5679–5688 (2017)
2. Hassner, T., Harel, S., Paz, E., Enbar, R.: Effective face frontalization in unconstrained images. In: Computer Vision and Pattern Recognition, pp. 4295–4304 (2014)
3. Hayat, M., Khan, S.H., Werghi, N., Goecke, R.: Joint registration and representation learning for unconstrained face identification. In: IEEE Conference on Computer Vision and Pattern Recognition, pp. 1551–1560 (2017)
4. Li, H., Brandt, J., Lin, Z., Shen, X., Hua, G.: A multi-level contextual model for person recognition in photo albums. In: Computer Vision and Pattern Recognition, pp. 1297–1305 (2016)

5. Liu, J., Wang, G., Hu, P., Duan, L.Y., Kot, A.C.: Global context-aware attention lstm networks for 3D action recognition. In: IEEE Conference on Computer Vision and Pattern Recognition, pp. 3671–3680 (2017)
6. Luan, T., Yin, X., Liu, X.: Disentangled representation learning GAN for pose-invariant face recognition. In: Computer Vision and Pattern Recognition, pp. 1283–1292 (2017)
7. Mahasseni, B., Lam, M., Todorovic, S.: Unsupervised video summarization with adversarial LSTM networks. In: Conference on Computer Vision and Pattern Recognition (2017)
8. Masi, I., Rawls, S., Medioni, G., Natarajan, P.: Pose-aware face recognition in the wild. In: Computer Vision and Pattern Recognition, pp. 4838–4846 (2016)
9. Moreno-Noguer, F.: 3D human pose estimation from a single image via distance matrix regression. In: Computer Vision and Pattern Recognition (2017)
10. Nech, A., Kemelmacher-Shlizerman, I.: Level playing field for million scale face recognition. In: Computer Vision and Pattern Recognition, pp. 3406–3415 (2017)
11. Oh, S.J., Benenson, R., Fritz, M., Schiele, B.: Person recognition in personal photo collections. In: IEEE International Conference on Computer Vision, pp. 3862–3870 (2015)
12. Rogez, G., Weinzaepfel, P., Schmid, C.: LCR-net: Localization-classification-regression for human pose. In: IEEE Conference on Computer Vision and Pattern Recognition, pp. 1216–1224 (2017)
13. Cheng, D., Gong, Y., Zhou, S., Wang, J., Zheng, N.: Person re-identification by multi-channel parts-based CNN with improved triplet loss function. In: IEEE Conference on Computer Vision and Pattern Recognition, pp. 1335–1344 (2016)
14. Li, X., Zheng, W.S., Wang, X., Xiang, T., Gong, S.: Multi-scale learning for low-resolution person re-identification. In: IEEE International Conference on Computer Vision, pp. 3765–3773 (2015)
15. Liao, S., Li, S.Z.: Efficient PSD constrained asymmetric metric learning for person re-identification. In: IEEE International Conference on Computer Vision, pp. 3685–3693 (2015)
16. Ren, S., He, K., Girshick, R., Sun, J.: Faster R-CNN: towards real-time object detection with region proposal networks. In: International Conference on Neural Information Processing Systems, pp. 91–99 (2015)
17. Farenzena, M., Bazzani, L., Perina, A., Murino, V., Cristani, M.: Person re-identification by symmetry-driven accumulation of local features. In: Computer Vision and Pattern Recognition, pp. 2360–2367 (2010)

Get the Whole Action Event by Action Stage Classification

Weiqi Li, Jianming Wang, Shengbei Wang, and Guanghao Jin[✉]

Department of Computer Science and Software Engineering,
Tianjin Polytechnic University, Tianjin, China
{wangjianming,jinguanghao}@tjpu.edu.cn

Abstract. Spatiotemporal action localization in videos is a challenging problem which is also an essential and important part of video understanding. Impressive progress has been reported in recent literature for action localization in videos, however, current state-of-the-art approaches haven't considered the scenario of broken actions, in which an action in an untrimmed video is not a continuous image series anymore because of occlusion, shot change, etc. So, one action is divided into two or more footages (sub-actions) and the existing methods localize each of them as an independent action. To overcome the limitation, we introduce two major developments. Firstly, we adopt a tube-based method to localize all sub-actions and discriminate them into three action stages with a CNN classifier: Start, Process and End. Secondly, we propose a scheme to link the sub-actions to a complete action. As a result, our system is not only capable of performing spatiotemporal action localization in an online-realtime style, but also can filter out irrelevant frames and integrate sub-actions into single tube that has better robustness than the existing method.

Keywords: Action stage classification · Action localization
Online real-time

1 Introduction

Video action localization [1–3] is one of the most important objectives of video understanding It can be applied in many real-life scenarios *e.g.* like video surveillance [13], video captioning [11,12]. In action localization case, appearance and motion are related and using both of them can lead to significant improvement in performance. Thus many methods use two-stream CNN approach [14,17,18] to deal with localization task. Generally, optical flow is used to present motion as

Supported by National Natural Science Foundation of China (Grant No. 61373104) and Natural Science Foundation of Tianjin (Grant No. 16JCYBJC42300 and Grant No. 17JCQNJC00100) and Science and Technology Commission of Tianjin Municipality (Grant Nos. 15JCYBJC16100) and Program for Innovative Research Team in University of Tianjin (No. TD13-5032).

K. Yoshida and M. Lee (Eds.): PKAW 2018, LNAI 11016, pp. 231–240, 2018.
https://doi.org/10.1007/978-3-319-97289-3_18

its invariance to image appearance. Then these methods get proposals extracted from the two stream by classifying and regressing the bounding boxes at the frame-level [1,4]. Finally, the frame-wise detection boxes are linked to form the action tubes [2,6] based on spatial overlapping rate and class-specific confidence score.

In online case [1,3], the existing methods starts generating tubes from the moment when the action is first detected. Then the tube ends when no action is detected over consecutive k frames. In some cases, the termination of a tube does not mean the end of the action. The figure below shows an example of this situation (see Fig. 1).

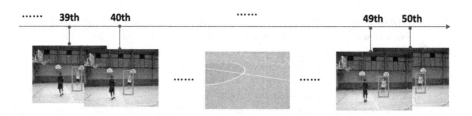

Fig. 1. This is a detection result when a shot change occurs. No action is detected due to the shot change after the 40th frames until the 49th frame, and the number of undetectable frames exceed the threshold, so the formed action tube is interrupted. However, the action is still in progress. From the 49th frame, a new tube is generated again because the camera switched back and detected the previously unfinished action. Those two sub-action tubes belong to a same action event.

Despite the success of exsiting methods, we noticed that those methods only focus on whether there is an action, what action it is, and where the action tube exists. They have not solve the problem of obtaining a complete action event from a bunch of scattered sub-action tubes. They do not understand that these parts originally belonged to a complete action event, not two separate actions. This leads to the lack of detecting continuity of the generated tubes and cannot be used to detect continuous behavior.

Paper [5,16] propose a temporal evolution framework which utilizes temporal structural information. It is based on the observation that all actions have three continuously components: "start", "middle", and "end". Each component has distinct patterns of appearance and motion.

In this work, we propose a new online real-time spatiotemporal action localization methods that can combine the scattered action tubes to a single tube for detecting a whole action event. Our model utilizes the concept of three components: "starting", "processing" and "ending" and can significantly improve localization performance. It merges scattered tubes by checking the stage of ending and starting parts of the tubes. Through the classification of the action tube stage, we can link two or more sub-action tubes by a link condition and filter the interference during an action event. We trained and evaluated our approach on the most challenging UCF101-24 dataset and achieve better performance than the existing methods.

2 Related Work

2.1 Off-line Methods

The existing off-line case assume that the video can be obtained in advance. Paper [2] utilizes the R-CNN to two-stream network: processing appearance and motion information separately. Two-stream Faster R-CNN has been introduced by [4, 7]. They use supervised RPNs to generate region proposal for actions on frame-level, and solved the S/T association problem via 2 recursive passes over frame level detections for the entire video by dynamic programming.

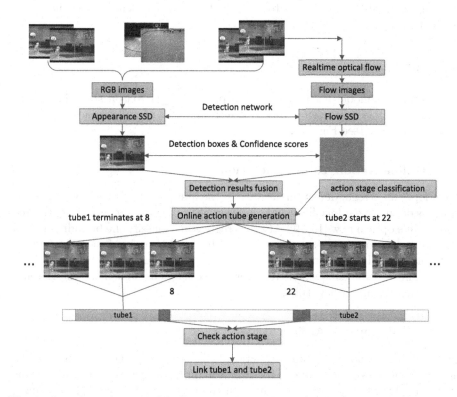

Fig. 2. Overview of our model. Given a sequences of video frames, we get the fusion detection boxes of every frame using two-stream detection network and link consecutive bounding boxes to form the preliminary tubes, then we need to discriminate the action stage of the broken part of the frames. When the action stage of two parts meet the link condition, we link them together to get a single tube for a whole action event. We can get a more complete action event by linking the two sub-action tubes.

2.2 On-line Methods

Paper [1] uses SSD detector to implement real-time case, it is proceeded by classifying and regressing a set of anchor boxes to the ground-truth bounding-box of the object to get predictions at the frame-level. Their linking algorithm works in an online fashion and tubes are updated frame by frame, together with overall action-specific scores and labels. Paper [3] is also based on the SSD framework and it leverages sequences of frames instead of operating on the frame-level and propose the action tubelet detector that proceeds by scoring and regressing anchor cuboids.

Those methods have good performance in on and off line localization cases [1, 3, 4]. In online case, we cannot get the whole view of an action as its appearance depends on the time steps. The scattered tubes may only contain partial parts of an action which makes it hard to detect the action type by using like deep learning methods. Thus, we try to solve that kind of problem by checking stage of the tubes.

3 Our Model

In general, our model need to solve two problems: (1) Action stage judgment; (2) Link multiple sub-action tubes. We introduce the overall framework for the above two issues (see Fig. 2).

3.1 Online Action Tube Generation

We adopt two-stream detection network as the first part of our framework, which consists of an appearance SSD and a motion SSD. The flow images are extracted by realtime optical flow [10, 19]. After we regress and classify the bounding boxes, the two boxes and action-specific score will be fused by union-fusion strategy [1] to form the final detection boxes. Then merge those consecutive frames to form tubes. Thus a tube ends when the interference happens. When the interference ended, there comes a new tube.

3.2 Classifying the Action Stage

In interference case, the obtained action tubes are likely the parts of a whole action event, sub-action tubes. To solve this problem, we try to use the time structure of the tubes which have three different stages: "starting", "processing", "ending". The action stage classification network classifies the ending part of former tube and the start part of following tube. It gives the confidence score of each stage, representing the possibility of being at that stage. Then we can analyze whether the current action's interruption is caused by external conditions or the end of its own action (see Fig. 3). We first trained a new network to detect the stages of the action for each frame. Each frame is labeled to "starting" or "processing" or "ending".

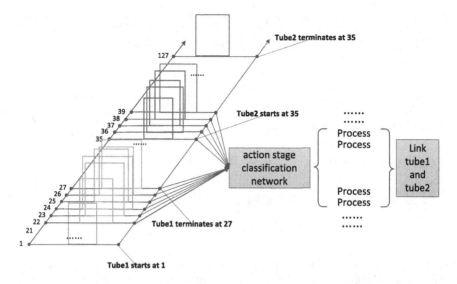

Fig. 3. The detection box of each frame is linked to generate the action tubes. We select last part of tube1 and start part of tube2 to check the action stage.

3.3 Link Conditions of Two Action Tubes

In addition to checking the action stage of broken part, we also introduce a threshold system to assist in the determination. The threshold refers to the number of interfering frames between two sub-action tubes belonging to the same action event. In [1], the tube terminates when no related action is detected more than 5 frames. In our work, we adjust the threshold to adapt different action stage and simulate real-world interference. There are 9 situations for the stage pair of these two parts.

In the case when the two parts are in the same stage (see Fig. 4), that is to say they are interrupted during a single stage, for example "starting-starting", As the starting and ending of a action will not continue long time, the threshold between those stage should be shorter than both "processing" case (case1 & case3).

In addition, when the current action tube terminates in an ending stage, we need to focus on the starting part of following tube. It can be link if the next tube starts with an "ending" stage and the number of unrelated frames between two parts does not exceed the threshold. Otherwise, we think that the current tube is terminated and the next new tube has started in the disturbed frames.

4 Implementation

The input is the fused frame-level detection boxes with their classification scores and stage of the action. At each time step t, for each coming frame, the top n class-specific detection boxes $\{b_c\}$ are picked by applying non maximum

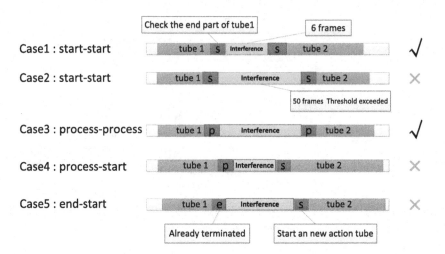

Fig. 4. Some different situations where the tube is interrupted. White part indicates the input video. Blue part refers to the detected tubes. Two greens indicates the action stage of the ending of the previous tube and the starting of the second tube respectively. The red part represents the broken part. At the same time we give the result of whether it can be linked. (Color figure online)

suppression on a per-class basis. In a frame the first time an action is detected, $n_c(1) = n$ action tubes per class are initialized using the n detection boxes at $t = 1$ ($t = 1$ indicates the start of the action tube) and the action stage is start in the first frame of the generated tube. The number of tubes $n_c(t)$ varies with time steps, as new tubes are added and/or old tubes are terminated. At each time step, we sort the existing partial tubes so that the best tube can potentially match the best box from the upcoming frame t. Also, for each partial tube τ_c^i at time $t-1$, we restrict the potential matches to detection boxes at time t whose IoU (Intersection over Union) with the last box of τ_c^i is above a threshold λ and tubes cannot simply drift off. As a summary, action tubes are constructed by applying the following 7 steps to every new frame at time t:

Step 1. Execute steps 2 to 6 for each class c.

Step 2. Sort the $n_c(t-1)$ action tubes generated up to time $t-1$ in decreasing order according to the mean of the class-specific scores of the tubes each detection boxes in every frame.

Step 3. LOOP START: Traverse the sorted tube list in Step 2 from $i = 1$ to $n_c(t-1)$.

Step 4. Select tube τ_c^i from the sorted list and find a matching box for it among the n class-specific detection boxes b_c^j, $j = 1,...,n$ at frame t based on the following conditions: (a) for all $j = 1,...,n$, if the IoU between the last box of tube τ_c^i and the detection box b_c^i is greater than, then add it to a potential match list β^i; (b) if the list of potential matches is not empty, $\beta^i = \phi$, select the box b^{max} from β^i with the highest score for class c as the match, and remove it from the set of available detection boxes at time t; (c) if $\beta^i = \phi$, check the stage of the last 20% of the current tube and record it, then retain the tube anyway and continue the detection of next frame until the next action tube show up. Check 20% of the beginning of the following tube and determine whether the two tubes satisfy the link conditions before and after.

Step 5. Update the temporal labelling for tube τ_c^i using the score $s(b^{max})$ of the selected box b^{max}.

Step 6. LOOP END

Step 7. If any detection box is left un assigned, start a new tube at time t using this box.

5   Experiments

5.1   Dataset:UCF-24

We choose the dataset which provides spatio-temporal ground-truth annotations. It is a subset of challenging UCF-101 [8] action dataset. It is composed of 24 sports action categories and 3200 videos. In addition, the videos are relatively longer and they are not temporally trimmed so as to localize the action instances in temporal. Each video only contains a single action category, on average there are 1.5 action instances per video or multiple people in a video perform the same action, with different spatial and temporal boundaries.

Training Data Construction: First of all, we label the remaining action tubes in the training set with three stages and found that most of the actions were divided into approximately 20%, 60%, 20%. We strip 500 videos from the training set that contain only a single action tube to the test set. After this operations, each divided part is labeled as "starting, processing, ending". We provide an analysis of our action stage classification network (Sect. 5.3).

5.2   Implementation Details

We adopt SSD [9] framework as our basic detection network and use ImageNet [20] pretrained VGG16 network [9] for both appearance and motion streams. The input RGB and flow frames are resized to 300×300 to get faster speed. As for action stage classification network, we use the LeNet [15] to get the classification results with processed dataset video for both appearance and motion streams.

5.3   Validation of Action Stage Classification

In this section we study the effectiveness of our action stage classification network. We select tubes from the test set consist of 300 videos and 500 tubes stripped from train set to form a new test set. The test set includes all 24 actions. We get three stages of scores by classification.

We verified the performance of the CNN action stage classification network on the new test set and obtained the results for the three stages of the classifier: starting, processing, ending. The final overall classification accuracy achieved 91.7 %.

Table 1. Three stages classification accuracy

Action stgae	Start	Process	End	Total
Accuracy(%)	92.93	92.16	89.90	91.7

5.4   Complete Action Tube Generation Performance

The acquisition of the action stage allowing us to determine whether the linking condition is satisfied. First, the stage of each end and start part of two separated tube is measured according to the stage of all frames in that part. We first select 20% of the tube and make a statistic on the stage of all the frames in this part. The stage with the most frames is selected as the stage of current part of tube.

Table 2. Double stage classification accuracy of truncated action tubes

Break tube stage	Double stage accuracy(%)
Start-start	90.49
Process-process	91.23
End-end	86.81
Start-process	90.19
Start-end	88.63
Process-end	87.13
End-start	97.63
Process-start	89.93
End-process	90.03

All 800 videos in test set possess a whole action instance. We truncate each tube in the test set from the middle to form sub-action tubes, so they are actually all linkable. After classification, we get 9 pairs of stage. According to the link condition, we get 604 pairs of action tubes that can be integrated. The results are shown in the above table (see Table 2).

Table 3. Link accuracy

Link result	Yes
Accuracy	90.13%

5.5   Comparison with the Existing Methods

As shown in Table 1, We compared our method with existing methods on the new dataset, with different IoU thresholds. The IoU defines the overlap between the detection box at time t and the last box of formed tube at time $t-1$. Table 4 reports the video-mAP comparison results at various IoU thresholds. After testing, our method can integrate the interrupted sub-actions into a single tube to achieve whole action event extraction. It shows that our method has made progress in detecting continuous action and filtering out irrelevant frames (Table 3).

Table 4. S/T action localisation performance(mAP) comparison on UCF101-24.

IoU threshold	0.2	0.5
Peng and Schmid [4]	57.44	10.65
Saha *et al.* [7]	50.82	14.33
Singh *et al.* [1]	55.37	21.05
SSD+stage classification	**71.49**	**39.95**

6   Conclusion and Future Works

In this paper, we presented a novel framework for spatiotemporal action localization that are able to identify the stage of the ongoing action. Different from existing methods, our method can extract the continuous behavior in the video, filter out unrelated frame interference, get the whole action event. Then based on deep learning strategy for the simultaneous detection and classification of region proposals with a new incremental action tube generation approach, our method achieves high accuracy and ensures fast detection speed.

References

1. Singh, G., Saha, S., Sapienza, M., Torr, P., Cuzzolin, F.: Online real-time multiple spatiotemporal action localisation and prediction. In: 2017 IEEE International Conference on Computer Vision (ICCV), pp. 3657–3666, October 2017. https://doi.org/10.1109/ICCV.2017.393
2. Gkioxari, G., Malik, J.: Finding action tubes. In: 2015 IEEE Conference on Computer Vision and Pattern Recognition (CVPR), pp. 759–768, June 2015. https://doi.org/10.1109/CVPR.2015.7298676
3. Kalogeiton, V., Weinzaepfel, P., Ferrari, V., Schmid, C.: Action tubelet detector for spatio-temporal action localization. In: 2017 IEEE International Conference on Computer Vision (ICCV), pp. 4415–4423, October 2017. https://doi.org/10.1109/ICCV.2017.472
4. Peng, X., Schmid, C.: Multi-region two-stream R-CNN for action detection. In: Leibe, B., Matas, J., Sebe, N., Welling, M. (eds.) ECCV 2016. LNCS, vol. 9908, pp. 744–759. Springer, Cham (2016). https://doi.org/10.1007/978-3-319-46493-0_45
5. Yuan, Z., Stroud, J.C., Lu, T., Deng, J.: Temporal action localization by structured maximal sums (2017)

6. Weinzaepfel, P., Harchaoui, Z., Schmid, C.: Learning to track for spatio-temporal action localization. In: 2015 IEEE International Conference on Computer Vision (ICCV), pp. 3164–3172, December 2015. https://doi.org/10.1109/ICCV.2015.362

7. Saha, S., Singh, G., Sapienza, M., Torr, P.H.S., Cuzzolin, F.: Deep learning for detecting multiple space-time action tubes in videos (2016)

8. Soomro, K., Zamir, A.R., Shah, M.: UCF101: a dataset of 101 human actions classes from videos in the wild. CoRR abs/1212.0402 (2012). http://arxiv.org/abs/1212.0402

9. Liu, W., Anguelov, D., Erhan, D., Szegedy, C., Reed, S.E., Fu, C., Berg, A.C.: SSD: single shot multibox detector. CoRR abs/1512.02325 (2015). http://arxiv.org/abs/1512.02325

10. Kroeger, T., Timofte, R., Dai, D., Gool, L.J.V.: Fast optical flow using dense inverse search. CoRR abs/1603.03590 (2016). http://arxiv.org/abs/1603.03590

11. Shen, Z., Li, J., Su, Z., Li, M., Chen, Y., Jiang, Y.G., Xue, X.: Weakly supervised dense video captioning. In: The IEEE Conference on Computer Vision and Pattern Recognition (CVPR), July 2017

12. Pasunuru, R., Bansal, M.: Multi-task video captioning with video and entailment generation, pp. 1273–1283 (2017)

13. Kaufman, D., Levi, G., Hassner, T., Wolf, L.: Temporal tessellation: a unified approach for video analysis. In: The IEEE International Conference on Computer Vision (ICCV), October 2017

14. Simonyan, K., Zisserman, A.: Two-stream convolutional networks for action recognition in videos. CoRR abs/1406.2199 (2014). http://arxiv.org/abs/1406.2199

15. Lecun, Y., Bottou, L., Bengio, Y., Haffner, P.: Gradient-based learning applied to document recognition. Proc. IEEE 86(11), 2278–2324 (1998). https://doi.org/10.1109/5.726791

16. Zhao, Y., Xiong, Y., Wang, L., Wu, Z., Lin, D., Tang, X.: Temporal action detection with structured segment networks. CoRR abs/1704.06228 (2017). http://arxiv.org/abs/1704.06228

17. Ren, S., He, K., Girshick, R., Sun, J.: Faster R-CNN: towards real-time object detection with region proposal networks. IEEE Trans. Pattern Anal. Mach. Intell. 39(6), 1137–1149 (2017). https://doi.org/10.1109/TPAMI.2016.2577031

18. Girshick, R.B., Donahue, J., Darrell, T., Malik, J.: Rich feature hierarchies for accurate object detection and semantic segmentation. CoRR abs/1311.2524 (2013). http://arxiv.org/abs/1311.2524

19. Zhang, B., Wang, L., Wang, Z., Qiao, Y., Wang, H.: Real-time action recognition with enhanced motion vector CNNs. CoRR abs/1604.07669 (2016). http://arxiv.org/abs/1604.07669

20. Krizhevsky, A., Sutskever, I., Hinton, G.E.: Imagenet classification with deep convolutional neural networks. In: International Conference on Neural Information Processing Systems, pp. 1097–1105 (2012)

Clothing Attribute Extraction Using Convolutional Neural Networks

Wonseok Lee, Sangmin Jo, Heejun Lee, Jungmin Kim,
Meejin Noh$^{(\boxtimes)}$, and Yang Sok Kim

Department of Mangement Information Systems, Keimyung University,
1095 Dalgubeol-daero, Dalseo-gu, Daegu 42061, South Korea
{mjnoh,yangsok.kim}@kmu.ac.kr

Abstract. Automated annotation of cloth images is an appealing technique, which have many applications, such as cloth search and classification. This study suggest an automated cloth attribute annotation system suggested that extracts low-level features from images and learns multiple classification models for predicting 25 cloth attributes. This study uses a deep convolutional neural networks (CNN) algorithm for building classifiers. Our research results show that CNN-based approach outperforms the previous approach.

Keywords: Image attribute extraction · Convolutional neural networks
Cloth attribute classification

1 Introduction

Nowadays images are one of the most significant data formats, but they usually do not provide semantic information. Finding the semantic information of images is crucial in many applications such as information retrieval and classification. In general, there are three approaches: First, humans can add semantic information by adding annotations to images. However, this approach is impractical if the number of images is large. Second, low-level semantic information, such as color, shape and texture, can be automatically obtained by analyzing images. However, some research shows that there is a semantic gap between low level semantic information and semantics by human. Third, high-level semantic information can be automatically extracted from the images using machine learning techniques [1]. This research mainly focuses on the second approach among these three approaches because this research focuses on finding similar clothes and on classifying clothes into defined categories.

Extracting attributes of clothing from natural images is extremely difficult because the appearance of clothing items are very different. Chen et al. [2] suggest a fully automated annotation system that generates semantic annotations for a given image. Broadly, their approach consists of four steps. First, the system extracts pose regions from natural images using a pose detection algorithm. Then, the system extracts low-level features from the detected pose by applying various feature extraction techniques and transform them into the combined features. Third, the system uses an SVM algorithm to predict cloth attributes. Finally, the system uses a conditional random field

© Springer Nature Switzerland AG 2018
K. Yoshida and M. Lee (Eds.): PKAW 2018, LNAI 11016, pp. 241–250, 2018.
https://doi.org/10.1007/978-3-319-97289-3_19

(CRF) approach in order to enhance the performance of attribute extraction results by considering mutual dependencies between attributes in the third step. Note that even though the final steps may improve attribute extraction performances, the third step is critical in the overall processes. For this reason, although we follow the approach suggested by [2], our research only focuses on the attribute extraction problem.

When Chen et al. extract attributes, they use classifiers modeled with a support vector machines (SVM) algorithm. Although they provided a fully automated annotation system with an SVM-based approach, it is not certain whether the SVM algorithm is the best for the problem. In this research, we employ deep convolutional neural networks (CNN) instead of SVM because it outperforms conventional machine learning algorithms in the object detection problem [3, 4]. Based on the cloth attribute extraction framework suggested by Chen et al., we compare performances of the CNN based classifier and the SVM based classifier.

2 Related Work

2.1 Semantic Annotation of Images

In the area of computer vision, a large amount of research has been conducted to extract semantic information that explains image objects, because they provide useful information for further processing. Manual semantic annotation supported by knowledge engineering and ontology technology, such as [5, 6], was initially suggested by researchers. However, manual annotation is inefficient and too subjective and thus is difficult to use as a viable solution [7]. Semi-automatic annotation approaches were suggested to overcome this limitation. In this approach, objects of images are identified and tagged with semantic information and then machine learning techniques are used to build models that are used to annotate unseen images [8, 9]. This approach for semantic annotation describes objects using their attributes.

2.2 Clothing Attributes Extraction from Images

Clothing attribute extraction is an example of semantic annotation. Yamaguchi et al. [10] suggest a clothing attribute extraction from fashion images. Their system detects segments, so called 'superpixels' of images [11], detects poses using an algorithm proposed by [12] and finally predicts their clothing labels by applying a conditional random fields (CRF) model. Chen et al. [2] suggest a clothing attribute extraction method that addresses the same problem addressed by [10]. A Major difference between [2] and [10] is that the former predicts low-level features using SVM classifiers and predicts the final attributes using the CRF model, while the latter directly predicts clothing attributes using the CRF model. This approach is categorized as type of pixel-wise labeling approach because a training set of images is annotated by a human supervisor. In this research, we only focus on the pixel-wise labeling approach and improving individual attribute prediction.

2.3   Clothing Attribute Extraction by CNN

CNN extends a standard multilayer neural network by combining one or more convolutional layers with one or more fully connected layers. This approach is designed to take advantage of the 2D structure of an input image and make it easier to train with fewer parameters than fully connected networks with the same number of hidden units. For this reason, several studies adopt CNN in clothing attribute extraction. Liang et al. [13] view the human parsing problem as an active template regression problem. They use CNN to build the end-to-end relationship between images and their attributes. Recently Liang et al. [14] suggest a newer method that uses contextualized CNN, which considers contexts of cross-layer, global image-level, semantic edge, within-super-pixel and cross-super-pixel neighborhood and its result outperforms that of Liang et al. [13]. However, these studies do not directly address clothing attribute extraction, instead address human parsing. Clothing parsing aims to obtain more fine-grained attributes in comparison to human parsing.

3   Method

3.1   Feature Extraction

It is difficult to extract clothing features from images because the humans in the images are in various poses. Chen et al. [2] extract clothing features by recognizing poses before extracting features from unstructured images. Chen et al. [2] recognize the pose by detecting face, upper body, and pose in sequence. Recently, Cao et al. [15] suggest a new pose estimation technique, referred to as Part Affinity Fields (PAFs). PAFs, which directly considers the association between anatomical parts in an image, outperforms previous state-of-the-art techniques. Therefore, we adopt this technique to identify poses and to extract low-level features in our research.

3.2   SIFT

Scale Invariant Feature Transform (SIFT) is a method for extracting invariant features from images regardless of their scale and rotation [16]. Essentially, this technique matches corresponding parts from two different images by extracting features characterized by SIFT and matching similar objects. Feature extraction processes using SIFT are summarized as follows: (1) The system generates blurred images through the Gaussian filter and image convolution calculation and constructs scale space. (2) The system makes the edges and corners present in the image stand out by applying Difference of Gaussian (DoG) between scale spaces of two blurring images. (3) The system finds key points from the images to which DoG have been applied (4) The system assigns one or more orientations based on gradient direction and scale information for each key point. (5) The system generates a feature vector, called the key point descriptor, consisting of 128 numbers that can represent characteristics of key points. In this research, we assigned 64×32 key points within the rectangle boundary of each image and extracted SIFT descriptors presented by 2048×128 matrix for each image and use them as features.

3.3   Texture Description

Texture description is useful when explaining the pattern of attributes of clothing. This research uses Local Binary Patterns (LBP) for extracting texture descriptors. It has been proven that the occurrence histograms generated by applying LBP to images are powerful texture feature [17]. In order to calculate LBP, the input image is transformed into grayscale, where the operator is invariant of any monotonic transformation of the gray scale. LBP is computationally simple because the operator can be realized with a few operations in a small neighborhood and a lookup table [18].

3.4   Color Description

Color is easily affected by environment. For example, a color can be differently viewed according to the level of light and degree of blurring. The RGB color model represents colors using red, green and blue colors and vary depending on the environment. Since RGB is easy to indicate a color in specific coordinate or pixel, it is often used to store images electrically or output images on the screen. However, in order to judge color, it do not have to be unaffected by the environment. To overcome this problem, this research used Hue-Saturation-Value (HSV) color model. HSV color space expresses color according to color, saturation, and brightness, and it is widely used for extracting an area of color. This research uses only H and S of HSV color space to minimize the effect of light. Since the images in our research use the RGB model, RGB values are converted into HSV values and H and S values are extracted as features.

3.5   Skin Probability

The degree of skin exposure varies widely depending on the type of clothing the people wear. For example, the skin exposure of a person wearing a tank top is higher than that of a person wearing a shirt. Thus, the degree of human skin exposure in an image can be an important feature in explaining clothing types and attributes. The skin detector applies an optimal threshold that distinguishes the skin from the color spaces of RGB, HSV, and YCbCr [19]. In this study, we use the skin ratio, which calculated by dividing the number of skin pixels detected by the skin detector with total numbers, as a feature.

3.6   Convolution Layer

Among deep learning algorithms, CNN exhibits high performance in image classification [20]. In CNN, after extracting the low-level-features in the convolutional layer, the high-level-features are extracted through a neural network. We use pose estimation to find the boundaries of the clothes, crop the image, resize it to 150 × 150, and extract the features of the image through the convolutional layer. The color classifier uses the HSV color model and extracts features with an image size of 150 × 150 × 3. The other classifier uses a gray scale and has an image size of 150 × 150 × 1 (Fig. 1).

Figure 2 shows our CNN structure as a whole. In convolutional layers of CNN structure, we apply batch normalization to all of convolutional layers. Max-pooling is

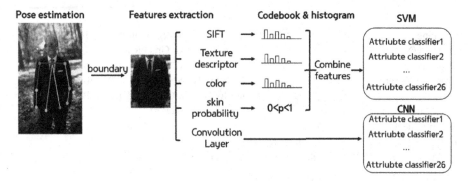

Fig. 1. Clothing attribute extraction process

used after each convolutional layer. A pooling layer which has 2×2 filter size and strides follows each convolutional layer which has 3×3 filter size. Subsequent to extracting low-level feature process, $7 \times 7 \times 64$ low features were extracted from the image through the convolutional filter and the pooling layer in the raw image.

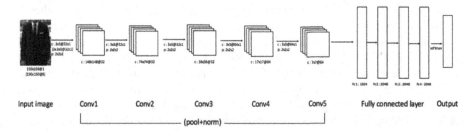

Fig. 2. CNN structure

3.7 Clothing Attribute Classifier

The classifier predicts clothing attribute labels summarized in Table 1, which are defined by Chen et al. [2]. The labels consist of 25 attributes in total, including 22 binary-class attributes as well as 3 multi-class attributes. This research compares the SVM-based classifier with the CNN-based classifier. After detecting pose, while the SVM classifier uses SIFT, texture descriptor, color descriptor and skin features, the CNN classifier generate features from the convolution layer.

The SVM-based Classifier. We adopt a Spatial Pyramid Matching (SPM) approach for the SVM-based classifier. SPM is a method considering the loss of geometric information in the Bag of Words (BoW). The BoW clusters the low-level features extracted from the images to find the code words that are the centers of the cluster. Then, a codebook composed of code words is generated for the entire image. Based on this codebook, each image is represented by a histogram composed of code words. This histogram value is interpreted as a feature vector and learned by SVM to distinguish the

Table 1. Clothing attributes

Clothing pattern (Positive/Negative)	Solid (1052/441), Floral (69/1649), Spotted (101/1619), Plaid (105/1635), Striped (140/1534), Graphics (110/1668)
Major color (Positive/Negative)	Red (93/1651), Yellow (67/1677), Green (83/1661), Cyan (90/1654), Blue (150/1594), Purple (77/1667), Brown (168/1576), White (466/1278), Gray (345/1399), Black (620/1124), >2 Colors(203/1541)
Wearing necktie	Yes 211, No 1528
Collar presence	Yes 895, No 567
Gender	Male 762, Female 1032
Wearing scarf	Yes 234, No 1432
Skin exposure	High 193, Low 1497
Placket presence	Yes 1159, No 624
Sleeve length	No sleeve (188), Short sleeve (323), Long sleeve (1270)
Neckline shape	V-shape (626), Round (465), Others (223)
Main category	Shirt (134), Sweater (88), T-shirt (108), Outerwear (220), Suit (232), Tank Top (62), Dress (260)

boundaries [21]. SPM does not obtain a histogram for the whole image; rather it gradually divides the image into sub-region and computing histograms of local features found inside each sub-region. Finally, a 'spatial pyramid' is formed by collecting all histograms. The histograms are built up for each sub-region, so that the geometric characteristics of the image can be reflected. In this study, a total of 1000, 256, 128 code words were obtained by applying k-Means for SIFT descriptor, texture descriptor, and color extracted from images. For SIFT descriptor, the feature vector was created by applying SPM up to level 1, and for texture descriptor and color, the feature vector was created by applying SPM up to level 0. Finally, the SVM model is learned by using a combined vector of all feature vectors and skin probability, and classified attributes describing clothes using it.

The CNN-based Classifier. The CNN-based classifier learns images through convolutions from raw images, unlike learning via SVM. The low-level features extracted from the convolutional layer are combined in the flat layer. In the fully connected layer, dropout is applied to nodes of all layers. The dropout deactivates a certain percentage of nodes, which can prevent overfitting of the neural network [22]. We applied a dropout of 0.8 for each layer and used 1024 nodes in hidden 1, 1024 nodes in hidden 2, 2,048 nodes in hidden 3, and 2048 nodes in hidden 4. In addition, initialization was done for all weights following the suggestion by Glorot and Bengio [23]. The error function in the output layer uses softmax. In order to reduce the bias in the course of learning, the sequence of images is mixed for every epoch and then it is learned by 30 batch sizes.

4 Experiments

Since the amount of data used in our experiments is not sufficient, we evaluate the performance using 5-fold-cross validation to increase the reliability of the performance measurement of the classifier. This evaluation was performed with the same data set for each classifier using SVM and CNN, and the performance was compared with that of Chen et al. [2]. When evaluating the performance of the classifier, the imbalance between the classes in the experimental data became a problem. Thus, we used the g-mean which is defined as

$$g - mean = \sqrt{recall \times precision}.$$

This measure is widely used as an indicator of the classifier for unbalanced data. For binary classifiers, positive values for recall and precision were set to Yes and measured the performance.

5 Results

Table 2 shows the average performance and standard deviation for 5-fold cross validation for each attribute. The cloth attribute was divided into clothing category, color, pattern and others. In Table 2, we can see that the attribute classification by the CNN-based classifier show a better g-mean in all attributes except color red and other collar than that of the SVM-based classifier. In particular, the CNN-based classifier shows 10% more than the SVM-Classifier when classifying necktie, neckline and scarf, graphics of spot and stripe pattern areas. The standard deviation of g-mean of the CNN-based classifier is less than those of the SVM-based classifier. This shows that the stability of classification performance of the CNN-based classifier is better than that of the SVM-based classifier.

Table 2. Clothing attributes

Attribute	SVM-based Classifier		CNN-based Classifier	
	g-mean	std	g-mean	std
(a) Clothing category				
Floarl	42.38%	13.12%	**44.46%**	5.97%
Plaid	34.54%	8.95%	**42.69%**	4.40%
Graphics	67.70%	4.11%	**79.60%**	3.59%
Solid	86.09%	2.13%	**93.51%**	0.55%
Spot	39.03%	8.37%	**60.35%**	5.55%
Stripe	46.30%	7.29%	**65.09%**	1.35%

(continued)

Table 2. (*continued*)

Attribute	SVM-based Classifier		CNN-based Classifier	
	g-mean	std	g-mean	std
(b) Color patterns				
Black	76.08%	2.66%	**80.52%**	1.07%
Yellow	50.87%	11.79%	**61.08%**	12.48%
Red	**63.24%**	11.98%	59.63%	4.45%
Cyan	55.67%	9.60%	**59.02%**	4.00%
Gray	57.30%	2.61%	**60.62%**	3.33%
Green	27.49%	2.96%	**60.15%**	4.92%
Brown	58.07%	2.96%	**63.72%**	2.63%
Blue	64.69%	5.20%	**67.03%**	4.43%
Purple	44.66%	9.13%	**54.51%**	11.12%
White	61.88%	5.12%	**68.08%**	2.75%
Black	76.08%	2.66%	**80.52%**	1.07%
(c) Others				
Category	52.71%	2.93%	**57.37%**	2.72%
Necktie	62.18%	5.18%	**80.75%**	2.01%
Neckline	54.18%	2.33%	**65.86%**	1.33%
Sleeve	83.01%	1.97%	**86.92%**	1.85%
Gender	80.07%	2.12%	**83.80%**	1.89%
Scarf	33.47%	5.27%	**59.64%**	6.36%
Placket	84.60%	1.55%	**86.25%**	1.17%
Collar	**84.91%**	1.36%	84.43%	1.10%

6 Conclusion

We propose a method for extracting semantic properties from clothing images. We can confirm that we have performance that conforms to unstructured images. In addition, we have confirmed that CNN that learns and classifies the image itself as an original feature shows higher performance and stability than the feature vector loading of SPM and SVM attribute classification, which have been widely used in existing computer vision. Chen et al. [2] increased the performance of classification results by adding CRF to the classification result by SVM. We can improve classification performance by applying CRF to the output nodes of the SVM-based and the CNN-based classifiers. Abdulnabi et al. [24] compared the performance of single-task CNN and multi-task CNN, and improved performance by applying Multi-task CNN. This suggests that the use of Multi - task CNN instead of CNN can improve classification performance. In the future we will improve the performance of the system for extracting clothing attributes by applying this research into our study.

References

1. Zhang, D., Islam, M.M., Lu, G.: A review on automatic image annotation techniques. Pattern Recogn. **45**(1), 346–362 (2012)
2. Chen, H., Gallagher, A., Girod, B.: Describing clothing by semantic attributes. In: Fitzgibbon, A., Lazebnik, S., Perona, P., Sato, Y., Schmid, C. (eds.) ECCV 2012. LNCS, vol. 7574, pp. 609–623. Springer, Heidelberg (2012). https://doi.org/10.1007/978-3-642-33712-3_44
3. Bappy, J.H., Roy-Chowdhury, A.K.: CNN based region proposals for efficient object detection. In: 2016 IEEE International Conference on Image Processing (ICIP) (2016)
4. Chi, Z., Li, H., Lu, H., Yang, M.H.: Dual Deep Network for Visual Tracking. IEEE Trans. Image Process. **26**(4), 2005–2015 (2017)
5. Hollink, L., Schreiber, A., Wielemaker, J., Wielinga, B.: Semantic annotation of image collections. in knowledge capture. In: 2003–Proceedings Knowledge Markup and Semantic Annotation Workshop (2003)
6. Erdmann, M., Maedche, A., Schnurr, H.-P., Staab, S.: From manual to semi-automatic semantic annotation: about ontology-based text annotation tools. In: Proceedings of the COLING-2000 Workshop on Semantic Annotation and Intelligent Content, pp. 79–85. Association for Computational Linguistics, Luxembourg (2000)
7. Little, S., Salvetti, O., Perner, P.: Semi-automatic semantic annotation of images. In: Seventh IEEE International Conference on Data Mining Workshops, ICDM Workshops. IEEE (2007)
8. Marques, O., Barman, N.: Semi-automatic semantic annotation of images using machine learning techniques. In: Fensel, D., Sycara, K., Mylopoulos, J. (eds.) ISWC 2003. LNCS, vol. 2870, pp. 550–565. Springer, Heidelberg (2003). https://doi.org/10.1007/978-3-540-39718-2_35
9. Farhadi, A., Endres, I., Hoiem, D., Forsyth, D.: Describing objects by their attributes. In: 2009 IEEE Conference on Computer Vision and Pattern Recognition (2009)
10. Yamaguchi, K., Kiapour, M.H., Ortiz, L.E., Berg, T.L.: Parsing clothing in fashion photographs. In: 2012 IEEE Conference on Computer Vision and Pattern Recognition (2012)
11. Arbelaez, P., Maire, M., Fowlkes, C., Malik, J.: Contour detection and hierarchical image segmentation. IEEE Trans. Pattern Anal. Mach. Intell. **33**(5), 898–916 (2011)
12. Yang, Y., Ramanan, D.: Articulated pose estimation with flexible mixtures-of-parts. In: CVPR 2011 (2011)
13. Liang, X., Liu, S., Shen, X., Yang, J., Liu, L., Dong, J., Lin, L., Yan, S.: Deep human parsing with active template regression. IEEE Trans. Pattern Anal. Mach. Intell. **37**(12), 2402–2414 (2015)
14. Liang, X., Xu, C., Shen, X., Yang, J., Tang, J., Lin, L., Yan, S.: Human parsing with contextualized convolutional neural network. IEEE Trans. Pattern Anal. Mach. Intell. **39**(1), 115–127 (2017)
15. Cao, Z., Simon, T., Wei, S.-E., Sheikh, Y.: Realtime multi-person 2D pose estimation using part affinity fields. In: Proceedings of the IEEE Conference on Computer Vision and Pattern Recognition (2017)
16. Lowe, D.G.: Distinctive image features from scale-invariant keypoints. Int. J. Comput. Vision **60**(2), 91–110 (2004)
17. Ojala, T., Pietikainen, M., Maenpaa, T.: Multiresolution gray-scale and rotation invariant texture classification with local binary patterns. IEEE Trans. Pattern Anal. Mach. Intell. **24**(7), 971–987 (2002)

18. Ojala, T., Pietikäinen, M., Mäenpää, T.: A generalized local binary pattern operator for multiresolution gray scale and rotation invariant texture classification. In: Singh, S., Murshed, N., Kropatsch, W. (eds.) ICAPR 2001. LNCS, vol. 2013, pp. 399–408. Springer, Heidelberg (2001). https://doi.org/10.1007/3-540-44732-6_41
19. Kolkur, S., Kalbande, D., Shimpi, P., Bapat, C., Jatakia, J.: Human skin detection using RGB, HSV and YCbCr color models. arXiv preprint arXiv:1708.02694 (2017)
20. Chatfield, K., Simonyan, K., Vedaldi, A., Zisserman, A.: Return of the devil in the details: delving deep into convolutional nets. arXiv preprint arXiv:1405.3531 (2014)
21. Csurka, G., Dance, C., Fan, L., Willamowski, J., Bray, C.: Visual categorization with bags of keypoints. In: Workshop on Statistical Learning in Computer Vision, ECCV 2004, Prague (2004)
22. Krizhevsky, A., Sutskever, I., Hinton, G.E.: ImageNet classification with deep convolutional neural networks. In: Proceedings of the 25th International Conference on Neural Information Processing Systems, vol. 12012, pp. 1097–1105. Curran Associates Inc., Lake Tahoe, Nevada (2012)
23. Glorot, X., Bengio, Y.: Understanding the difficulty of training deep feedforward neural networks. In: Yee Whye, T., Mike, T. (eds.) Proceedings of the Thirteenth International Conference on Artificial Intelligence and Statistics, PMLR Proceedings of Machine Learning Research, pp. 249–256 (2010)
24. Abdulnabi, A.H., Wang, G., Lu, J., Jia, K.: Multi-task CNN model for attribute prediction. IEEE Trans. Multimedia 17(11), 1949–1959 (2015)

Research Paper Recommender Systems on Big Scholarly Data

Tsung Teng Chen[1] and Maria Lee[2(✉)]

[1] National Taipei University, New Taipei City, Taiwan
timchen.ntpu@msa.hinet.net
[2] Shih Chien University, Taipei, Taiwan
Maria.lee@g2.usc.edu.tw

Abstract. Rapidly growing scholarly data has been coined Big Scholarly Data (BSD), which includes hundreds of millions of authors, papers, citations, and other scholarly information. The effective utilization of BSD may expedite various research-related activities, which include research management, collaborator discovery, expert finding and recommender systems. Research paper recommender systems using smaller datasets have been studied with inconclusive results in the past. To facilitate research to tackle the BSD challenge, we built an analytic platform and developed a research paper recommender system. The recommender system may help researchers find research papers closely matching their interests. The system is not only capable of recommending proper papers to individuals based on his/her profile, but also able to recommend papers for a research field using the aggregated profiles of researchers in the research field.

The BSD analytic platform is hosted on a computer cluster running data center operating system and initiated its data using Microsoft Academic Graph (MAG) dataset, which includes citation information from more than 126 million academic articles and over 528 million citation relationships between these articles. The research paper recommender system was implemented using Scala programming language and algorithms supplemented by Spark Mlib. The performance of the recommender system is evaluated by the recall rate of the Top-N recommendations. The recall rates fall in the range of 0.3 to 0.6. Our recommender system currently bears the same limitation as other systems that are based on user-based collaborative filtering mechanisms. The cold-start problem can be mitigated by supplementing it with the item-based collaborative filtering mechanism.

Keywords: Big Scholarly Data · Recommender systems
Research paper recommender systems · Collaborative filtering

1 Introduction

Recommender Systems are software systems and techniques that suggest items to a user based on predicted user preference rating. As a subclass of information filtering systems, it tries to predict the "preference" or "rating" a user would give to an item. To tame the information explosion, recommender systems have become increasingly popular in

© Springer Nature Switzerland AG 2018
K. Yoshida and M. Lee (Eds.): PKAW 2018, LNAI 11016, pp. 251–260, 2018.
https://doi.org/10.1007/978-3-319-97289-3_20

recent years, and are applied in a variety of areas including entertainment content such as movies and music, knowledge and information acquirement such as news and research articles, and purchase suggestion for products in general. As a branch of recommender systems study, research paper recommender systems are more in the spotlight partially due to the already enormous and still fast growing research-related information available online. A research paper recommender system aims to mitigate the information overload and helps scholars to find relevant research papers suited to their interests. A research scholar may need to locate relevant papers to keep track in his or her field of study or to cite articles pertinent to an article s/he is working on.

Based on a literature survey study, at least 217 articles relevant to research paper recommendations had been published by 2013. About 120 different recommendation approaches were discussed in these articles. The recommendation approaches were categorized into seven main classes – Stereotyping, Content-based Filtering (CBF), Collaborative Filtering (CF), Co-Occurrence, Graph-based, Global Relevance, and Hybrid [1]. The Stereotyping approach was inspired by subject stereotyping found in the field of psychology that provides a mechanism of quickly judging people based on a few personal characteristics [2]. For example, a typical stereotype would be "woman is more interested in romance than man". Stereotypes could be constructed through a collection of personal traits and then applied in a recommendation setting. Content-based Filtering and Collaborative filtering are widely utilized recommendation mechanisms in various applications. CBF infers users' interests profile from the items the users interacted with, whereas an item is modeled and represented by its features. In the context of research paper recommender systems, word-based features are commonly used. Features of a paper are extracted from its textual content. The similarity of papers is calculated by comparing their features and used subsequently by the recommendation mechanism to recommend a paper that is similar to what the users like. CF tries to find like-minded users by preference ratings given by them. Two users are considered like-minded if they rate items alike. With the pool of identified like-minded users, items that rated positively by a user become recommending candidates for other users in the pool. The co-occurrence recommendation approach refers to the practice of recommending related items to a user. The relatedness between items may be established by items' co-occurrence, such as two papers are both cited by another paper, which creates a co-citation relationship between the two cited papers. The graph-based approach abstracts various relationships between entities into a graph and applies graph metrics, such as distance and centrality, to find recommendation candidates. The relationships used in a graph-based approach may include a citation or co-citation relationship between papers, co-authorship between authors, or venues of papers etc. The global relevance approach decides recommendation by utilizing some global metrics, such as the citation counts or the h-index of a publication or an author. The h-index is a metric that attempts to measure the impact of an author or a scholarly journal. The hybrid recommendation approach refers to combining two or more aforementioned methods into one. Despite various approaches that have been proposed, it remains unclear which one is more promising for many reasons [1]. One of the main reasons is dataset discrepancy, which refers to different datasets or different versions of a dataset that are used in the studies. Direct comparison between performance metrics calculated from different approaches are problematic since datasets may critically influence the performance of a recommender system. The scholarly datasets of research paper recommender systems have grown in size recently and are referred to as big scholarly data, which have been discussed in several articles [3–5]. The research paper recommender system is among one of the main applications in the analytics of big scholarly data.

We tried to achieve several objectives in this study. One was to build a research paper recommender system utilizing recommendation mechanisms architected by open source projects. Another objective was to use publicly available big scholarly datasets to have a common basis. We hope to make the study of research paper recommender systems reproducible by utilizing publicly available architectures and datasets.

2 Current Status of Research Paper Recommender Systems

2.1 Research Paper Recommender Systems Related Studies

At least 217 research paper recommender systems related papers had been published by 2013 [1]. The main drawbacks of these researches are the unreproducible and incomparable results. The problems of reproducibility and comparability are due to several commonly found issues. The foremost issue is different datasets are used in the recommender systems that make the comparison between studies impractical. The datasets or data sources commonly used include CiteSeer and CiteULike, which account for 43% of all studies reviewed [1]. Most of the other datasets are taken from data sources that are often not publicly available. Another issue is that only a few papers disclose the architecture of their recommender systems. Two architectures for academic information collecting and pre-processing were discussed – system architecture for retrieval papers' PDF files by CiteSeer and the architecture for aggregating data usage from multiple academic data sources [1]. Another study describes an architecture platform that is capable of harvesting big scholarly information and hosting related applications such as citation recommendation [3]. Some recently published research paper recommender systems related articles utilized the afore-mentioned Graph-based and Global Relevance approaches [6] or used the collaborative filtering approach [7].

2.2 Research Paper Recommender Systems from the Perspective of Big Scholarly Data

Big scholarly data may be utilized in literature (research papers') recommendation, collaboration recommendation, and venue recommendation [4, 5]. An architecture platform tailored for big scholarly data analytics has been explored by the CiteSeer research team [3]. A recommender system utilizing Hadoop and Apache Mahout was introduced in a digital library recommender system [8], which makes recommendations based on roughly 2.2 million publications extracted from DBLP dataset.

2.3   Public Available Big Scholarly Datasets

As discussed earlier, the size of datasets used in previous research paper recommender studies ranges from ten thousand to a few million [1, 7–10]. In KDD Cup 2016, Microsoft granted Microsoft Academic Search (MAS) [11] dataset to be used freely in the KDD competition. The MAS dataset includes the Meta information of 126 million academic papers, 114 million authors, and over 528 million citations relationship between these papers. Our recommender system is built on the MAS dataset whose schema is shown in Fig. 1. Open Academic Society (OAS) [12] also has archived a more recent copy of MAS dataset, which includes bibliographical information of over 166 million academic papers. OAS also archived the A Miner dataset, which includes information on more than 154 million academic papers. Semantic scholar, which is funded by Microsoft cofounder Paul Allen, also has made their 20 million+ bibliographical dataset publicly available.

File Name	Fields	Size	Data Size
Papers	Paper ID Original paper title Normalized paper title Paper publish year Paper publish date Document Object Identifier (DOI)	27.2GB	126,909,022 Papers
Authors	Author ID Author name	2.66GB	114,698,045 Authors
Conferences	Conference ID Conference name	79KB	1,283 Conferences
Journals	Journal ID Journal name	972KB	23,404 Journals
PaperKeywords	Paper ID Keyword name Field of study ID mapped to keyword	4.99GB	158,280,967 Keywords
PaperReferences	Paper ID Paper reference ID	9.35GB	528,682,290 Paper References
FieldsOfStudy	Field of study ID Field of study name	1.43MB	53,834 Research Fields

Fig. 1. Partial schema of the MAS dataset

3 Research Paper Recommender Systems for Big Scholarly Data

3.1 The Author/Paper Utility Matrix for Recommender Systems

We utilized the widely used CF mechanism in our recommender systems for several reasons. Firstly, it is a mechanism that has been implemented on many platforms, including the Spark's scalable machine learning library (MLib), which was adopted by us. CF mechanism requires the interaction data between users and items to make recommendations. The traditional interaction data between users and items in a CF-based system are the explicit rating scores given to items by users. However, the implicit ratings given by users are usually infrequent and sparse, making the CF mechanism inoperative in some circumstances. To mitigate the problem of data sparsity, implicit interaction data between users and items are utilized. The implicit interaction data generally refers traces of data left unconsciously by users when they interact with items, such as web browsing logs or purchasing records [13]. In the context of research paper recommender systems, we postulate the behavior of citing or referencing academic papers approximates the explicit rating behavior. A citation is a conscious action made by an author. However, when an article is cited multiple times, it does not necessarily mean the article is regarded highly by an author. The main motivations of citing a paper were categorized as: (1) Perfunctory- an acknowledgement of some other relevant works have been performed; (2) Organic- facilitating the understanding of the citing article; (3) Conceptual- connecting a concept or theory that is used in the citing article; (4) Operational- referring the tools and techniques used in the citing article; (5) Evolutionary- the citing articles built on the foundations provided by the cited article [14]. It is fair to say that a cited article provides some utility to an author just like the enjoyment utility an entertaining item (e.g., a movie) to a viewer. Although the motivation for citing an article may differ, the aggregated citation count recorded by a paper is still regarded as a reliable measure of academic impact [15]. In line with this, we take the accumulated citation counts to an article as the proxy of the preference rating of an item. A higher citation count is equivalent to a higher preference rating. Analogous of a user/item preference rating matrix required by the CF recommender algorithm, an author/paper utility matrix is built, whereas authors as rows and articles as columns entries, respectively. The citation count an article received from an author is listed in the corresponding preference entry in the matrix. A simplified author/paper utility matrix is shown in Fig. 2. The paper-citation bibliographical data also has been utilized differently in other studies. For instance, a paper is regarded as a user and a citation is treated as an item in several studies to construct a paper-citation relations matrix [7, 16] for CF processing. However, instead of the more informative citation counts, only a binary relationship (a paper is cited or not) could be represented by this approach. Another study built the recommender mechanism using some graph-based operations over the citation network, which is derived from the paper-citation data [10].

	Paper 1	Paper 2	Paper 3	Paper 4	...	Paper m
Author 1		5	1	7	...	
Author 2	1		1	6	...	1
Author 3		1	8	2	...	1
Author 4	1		1	1	...	
...	...	...	...	...	...	...
Author n			5	4	...	

Fig. 2. An Author/Paper Utility Matrix. The citations count a paper received from an author stored in a cell in the matrix. The citation count is obtained by summing the total number of times a paper is cited by an author. Empty cells indicate no citation received.

3.2   The Architecture of BSD Capable Recommender Systems

Since the already massive scholarly data is expected to grow at an even faster pace, the capability of processing large dataset is now essential for recommender systems. The proposed platform should be capable of hosting massive and fast accumulating data, and supplementing mechanisms to facilitate efficient BSD analytics. In light of the considerations above, the Berkeley Data Analysis Stack (BDAS) [17] was selected as the main constituent for our BSD recommender systems. The BDAS stack includes a computer cluster manager (Mesos) that enables efficient resource virtualization and sharing across distributed applications and frameworks. Mesos supports Hadoop, Spark, and other applications through a dynamically shared pool of computing and storage resources. The architecture of our BSD-based recommender systems is shown in Fig. 3. The MAS dataset is parsed from its original text format and stored in HDFS format. The recommender system is implemented in Scala programming language utilizing the Alternating Least Squares (ALS) [18] algorithm provided by the Spark MLib. The data in the author/paper utility matrix are divided into 80/20% for training and test data, respectively. ALS is then applied to the training data iteratively to derive the low-rank matrices combination that have a minimum Root Mean Square Error (RMSE). The RMSE of test data is then calculated by applying the resulting low-rank matrices. The RMSE values computed from the training and test data are compared to check if an overfitting occurred. We may adjust the regularization hyper-parameter lambda and rerun ALS to fix the overfitting problem. To recommend research papers for a designate scholar, we just need to locate his corresponding row in the ALS-processed author/paper utility matrix. The values in the selected row correspond to the utility/preference rankings of the scholar. From this row, we then choose N entries with the highest values, which correspond to N highest-ranked candidate papers for Top-N [19] recommendation. The Top-N recommendations for a research field are obtained by summing the values from columns corresponding to papers in the research field from the utility matrix. The research field attribute of a paper is derived from the PaperKeywords and FieldOfStudy files in the MAS dataset.

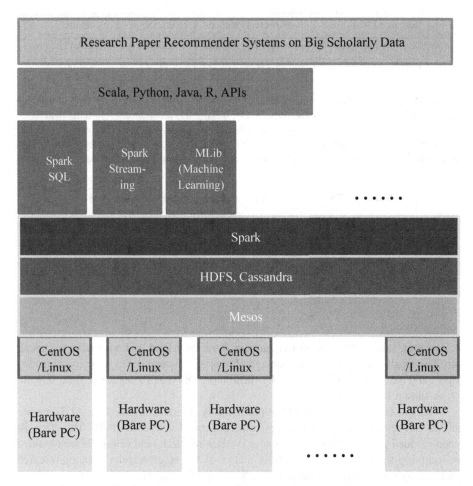

Fig. 3. The architecture of the recommender systems

4 Performance Evaluation of the Research Paper Recommender Systems

We evaluate the system performance using the offline metrics – recall. Since recall only considers the positively rated articles (cited articles in our case) within the Top-N, a high recall rate with lower N signified a better system [20]. For each author, the recall is calculated as follows:

$$recall = \frac{\text{number of articles the author cited in TopN}}{\text{total number of articles the author cited}}$$

We extracted datasets for two research fields from the MAS dataset, the information on the two datasets are shown in Fig. 4.

Research Field	papers	authors	references	# of citations
Machine Learning	27,117	53,712	230,552	1,630,572
Recommender Systems	5,431	10,281	32,545	452,667

Fig. 4. The paper column stores the number of papers published in the research field of machine learning and recommender systems, respectively. Taking the recommender systems field as an example, it includes 5,431 papers that were authored or co-authored by 10,281 distinct scholars and contained 32,545 references. There are 452,667 citations recorded between authors and papers (including papers' references) in the recommender systems research field.

Table 1. The recall rate of recommender systems research field

Author ID	Number of hits	Recall rate
76666523	7	0.35
72694593	12	0.6
77527215	6	0.3
76545162	7	0.35
71946686	8	0.4

We then randomly selected five authors from each field and calculated the recall rate. The recall rate here is the percentage of overlap between the top 20 recommended papers and the 20 most cited papers by the author. The recall rates range from 0.3 and 0.6 as shown in Table 1.

With the completely filled author/paper utility matrix computed by ALS, we are able to find the Top-N most recommended papers of a research field. The 20 most recommended papers in the recommender systems research field (Table 2) are obtained by summing the columns of the utility matrix of the recommender systems research field and retrieving the 20 columns with the highest summation values.

Table 2. The Top 20 papers in the recommender systems research field

Paper ID	Paper Title
10C9E0EA	Hybrid Recommender Systems: Survey and Experiments
7A283611	Latent Semantic Models for Collaborative Filtering
7A6FB77C	Matrix Factorization Techniques for Recommender Systems
7C54E0A8	An Algorithmic Framework for Performing Collaborative Filtering
7DC9036C	Empirical Analysis of Predictive Algorithms for Collaborative Filtering
7EA2B2D5	Social Information Filtering: Algorithms for Automating Word of Mouth
7537398E	Using Collaborative Filtering to Weave an Information Tapestry
757BB126	Evaluating Collaborative Filtering Recommender Systems
7FAE89BB	Item-based Top- N Recommendation Algorithms
7F3B2BC5	Explaining Collaborative Filtering Recommendations

(*continued*)

Table 2. (*continued*)

Paper ID	Paper Title
7F6B27CB	A Framework for Collaborative, Content-Based and Demographic Filtering
76FCDFDA	Analysis of Recommendation Algorithms for E-commerce
77270A42	GroupLens: Applying Collaborative Filtering to Usenet News
79018AC7	Recommending and Evaluating Choices in a Virtual Community of Use
79BABCCB	Item-based Collaborative Filtering Recommendation Algorithms
79CBDC59	Fab: Content-based, Collaborative Recommendation
80A853E8	Methods and Metrics for Cold-start Recommendations
80B12C04	Amazon.com Recommendations: Item-to-item Collaborative Filtering
80745098	GroupLens: An Open Architecture for Collaborative Filtering of netnews
81757DC2	Toward the Next Generation of Recommender Systems: A Survey of the State-of-the-art and Possible Extensions

5 Conclusion

We demonstrated the feasibility of developing a BSD-capable recommender system that is capable of making personalized recommendations. We may recommend the Top-N papers for an author based on his profile, which is derived from references listed in the articles authored by him or her. In addition, we are able to recommend the Top-N papers in a research field based on the aggregated authors' profiles in the field. The study could not have been done without the MAS dataset contributed by Microsoft. The rich meta-information included in the MAS dataset has made feasible many previously unthinkable analyses. Instead of painstakingly harvesting the vast scholarly data ourselves (as seen in many previous studies), academia should vigorously utilize the vast and rich datasets donated by the industry. We could better use our time and effort to develop novel applications from the publicly available big scholarly datasets compiled by the Open Academic Society or Semantic Scholar.

References

1. Beel, J., et al.: Research-paper recommender systems: a literature survey. Int. J. Digit. Libr. **17**(4), 305–338 (2016)
2. Rich, E.: User modeling via stereotypes. Cogn. Sci. **3**(4), 329–354 (1979)
3. Wu, Z., et al.: Towards building a scholarly big data platform: challenges, lessons and opportunities. In: Proceedings of the 14th ACM/IEEE-CS Joint Conference on Digital Libraries, London, United Kingdom, pp. 117–126. IEEE Press (2014)
4. Khan, S., et al.: A survey on scholarly data: From big data perspective. Inf. Process. Manag. **53**(4), 923–944 (2017)
5. Xia, F., et al.: Big scholarly data: a survey. IEEE Trans. Big Data **3**(1), 18–35 (2017)
6. Sesagiri Raamkumar, A., Foo, S., Pang, N.: Using author-specified keywords in building an initial reading list of research papers in scientific paper retrieval and recommender systems. Inf. Process. Manag. **53**(3), 577–594 (2017)

7. Haruna, K., et al.: A collaborative approach for research paper recommender system. PLoS ONE **12**(10), e0184516 (2017)

8. Ismail, A.S., Al-Feel, H.: Digital library recommender system on Hadoop. In: Proceedings of the 2015 IEEE 4th Symposium on Network Cloud Computing and Applications, pp. 111–114. IEEE Computer Society (2015)

9. Xia, F., et al.: Scientific article recommendation: exploiting common author relations and historical preferences. IEEE Trans. Big Data **2**(2), 101–112 (2016)

10. Son, J., Kim, S.B.: Academic paper recommender system using multilevel simultaneous citation networks. Decis. Support Syst. **105**, 24–33 (2018)

11. Sinha, A., et al.: An overview of microsoft academic service (MAS) and applications. In: Proceedings of the 24th International Conference on World Wide Web, Florence, Italy, pp. 243–246. ACM (2015)

12. Open Academic Society (2017). https://www.openacademic.ai/. Accessed 3 Jan 2018

13. Hu, Y., Koren, Y., Volinsky, C.: Collaborative filtering for implicit feedback datasets. In: Proceedings of the 2008 Eighth IEEE International Conference on Data Mining, pp. 263–272. IEEE Computer Society (2008)

14. Cano, V.: Citation behavior: classification, utility, and location. J. Am. Soc. Inf. Sci. **40**(4), 284–290 (1989)

15. Lutz, B., Hans-Dieter, D.: What do citation counts measure? A review of studies on citing behavior. J. Doc. **64**(1), 45–80 (2008)

16. McNee, S.M., et al.: On the recommending of citations for research papers. In: Proceedings of the 2002 ACM Conference on Computer Supported Cooperative Work, New Orleans, Louisiana, USA, pp. 116–125. ACM (2002)

17. Singh, D., Reddy, C.K.: A survey on platforms for big data analytics. J. Big Data **2**(1), 8 (2014)

18. Zachariah, D., et al.: Alternating least-squares for low-rank matrix reconstruction. IEEE Signal Process. Lett. **19**(4), 231–234 (2012)

19. Cremonesi, P., Koren, Y., Turrin, R.: Performance of recommender algorithms on top-n recommendation tasks. In: Proceedings of the Fourth ACM Conference on Recommender Systems. ACM (2010)

20. Wang, C., Blei, D.M.: Collaborative topic modeling for recommending scientific articles. In: Proceedings of the 17th ACM SIGKDD International Conference on Knowledge Discovery and Data Mining, San Diego, California, USA, pp. 448–456. ACM (2011)

Classification of CSR Using Latent Dirichlet Allocation and Analysis of the Relationship Between CSR and Corporate Value

Kazuya Uekado$^{(\boxtimes)}$, Ling Feng, Masaaki Suzuki, and Hayato Ohwada

Department of Industrial Administraion, Tokyo University of Science, 2641,
Yamazaki, Noda, Chiba, Japan
7417602@ed.tus.ac.jp, {fengl,m-suzuki,ohwada}@rs.tus.ac.jp

Abstract. Corporate social responsibility (CSR) is a business approach that aims to help address social or environmental problems. Many
researchers conducted empirical research to identify the relationship
between CSR activities and corporate value. Some researchers explain
that CSR is positively correlated with corporate value, others explain
that CSR is negatively correlated with corporate value. This disagreement among the researchers has arisen because CSR standards are
ambiguous. Therefore, we use topic classification to create a CSR standard. We rank the CSR activities. Our approach involves two steps.
First, a CSR standard is constructed using a topic model from CSR
reports. Second, the CSR rankings are calculated by using a random forest to calculate the importance of features related to CSR activities. The
results show a new CSR standard. Topics represents activities related to
reducing CO_2 emissions or diversity promotion, however it is not helpful
to consider too many topics: with more topics, more unrelated topics
appear. CSR rankings show that medical activities have the strongest
relationship with corporate value.

Keywords: Data mining · Topic classification
Latent Diriclet allocation · Machine learning
Corporate social responsibility

1 Introduction

1.1 Background

Social problems such as globalization and the environment, have recently been
growing. Corporate social responsibility (CSR) is a business approach that aims
to help address these problems. Since stakeholders expect companies to solve
such problems. Therefore, CSR is an important business strategy for ensuring
the sustainability of companies. CSR activities such as reducing carbon foot-
prints, volunteer activity for community or protecting human rights, are mainly

© Springer Nature Switzerland AG 2018
K. Yoshida and M. Lee (Eds.): PKAW 2018, LNAI 11016, pp. 261–270, 2018.
https://doi.org/10.1007/978-3-319-97289-3_21

conducted by large companies, who increase corporate value by releasing documentary evidence of their CSR processes and results.

In the academic world, many researchers are very interested in the relationship between CSR activities and corporate value and have conducted empirical research to identify such relationships.

In the 1970s, institutions around the US and northern Europe led the CSR researches. Orlitzky et al. [1] conducted a meta-analysis of previous studies, finding that CSR is positively correlated with corporate value. However, other researchers have found a negative relationship between CSR and corporate value.

This disagreement among the researchers has arisen because the definition of CSR is ambiguous, leading to differing views and a lack of comparability.

There are now CSR frameworks that capture CSR as it pertains to companies' ethical, social, and environmental performance. The first was defined in 2000 when several institutions came together to publish an international CSR standard, aiming to encourage companies to participate in CSR by providing clear definitions and useful guidelines. There are now several international CSR standards, including ISO 26000, GRI, and UNGC.

Based on ISO 2600, Maki and Feng [2] used machine learning (specifically, random forests) to compare the CSR efforts of different companies. They observed the changes in companies' CSR activities and ranked them by importance. However, we believe that using an international CSR standard such as ISO 26000 was inappropriate, for two reasons. First, such standards cannot follow CSR trends. Companies' CSR activities are continually changing as social problems change, and new initiatives may fall outside the standard's CSR framework. Second, CSR evaluations are necessarily subjective, even though CSR activities must be converted into numerical features for use in machine learning, to evaluate them based on ISO 26000. This means that the results are subjective and the CSR evaluation depends on the evaluator. If the evaluator changes, the results may change as well. Therefore, studies should not use international CSR standards.

1.2   Purpose

Based on Maki and Feng's research, we rank CSR activities in terms of their contribution to corporate value. We use topic classification to create a CSR standard, instead of relying on any existing international standard. We then use a random forest to rank the CSR activities. The key aim of this research is to evaluate CSR objectively, not subjectively, using an algorithmic process.

2   Data

This section introduces the data used in this research.

2.1 Target Companies

We used two types of data, one based on Nikkei NEEDS and another based on CSR reports. To compare our results with those of Maki and Feng, we limited the target companies to those satisfying the following four conditions.

1. Manufacturers listed in the first section of the Tokyo Stock Exchange
   CSR activities are mainly conducted by large companies. The first section of the Tokyo Stock Exchange is the highest market rank in Japan, and has strict listing criteria. We limited our attention to companies listed in this section so as to focus on large companies. It is also difficult to compare different industries, for two reasons. First, corporate values vary depending on the industry because earnings structures vary. Second, CSR activities also vary depending on the industry: manufacturing companies focus on the environment, while other companies focus on different types of CSR activity. We therefore limited our research to one industry, and chose manufacturing companies because they conduct a variety of CSR activities.
2. Companies that publish CSR reports
   This research creates CSR standards from CSR reports. Since we planned to create CSR standards based on CSR reports, we targeted companies that issue annual CSR reports and post them on their home pages.
3. Companies listed in the Nikkei NEEDS CD-ROM
   Nikkei NEEDS is a data service managed by Nikkei Inc. We planned to calculate the corporate values by using the performance data given by the Nikkei NEEDS CD-ROM as labels, and so required companies to be listed in Nikkei NEEDS.
4. Companies disclosing financial data with fiscal years ending in March
   Corporate values depend on the current social conditions, which can change due to various factors: either the Government implements a new regulatory law, or a potential disaster occurs. In addition, we focus on companies with fiscal years ending in March because this is common in Japan.

Compared with Maki and Feng's research, we added a new condition (Condition 2) requiring companies to produce CSR reports. A total of 167 companies satisfied the above conditions.

2.2 CSR Reports

CSR reports are brochures that companies publish about their CSR activities annually. To appeal to stakeholders, they typically post these on their home pages in PDF format. They contain company profiles, recent achievements, and details of the company's CSR activities. Such reports can also be called, for example, annual, integrated, environmental, or sustainability reports, but we call all such reports CSR reports.

The CSR library.net [3] site, managed by Brains Network Inc., collects CSR reports. As if April 2018, it lists 661 companies' reports. Users can view and download the reports by searching for a company or report name. We crawled CSR library.net to gather the target companies' CSR reports. For companies that were not listed, we collected the reports manually from their home pages.

2.3 Nikkei NEEDS

Nikkei NEEDS [4] is a data service managed by Nikkei Inc. It records performance data such as net sales, gross profit and capital stock, for companies listed in the first section of the Tokyo Stock Exchange. In this study, we used Nikkei NEEDS to calculate corporate values.

3 Method

3.1 Overview

Figure 1 shows an overview of our method.

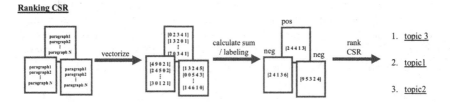

Fig. 1. Method overview

Our approach involves three steps. First, the data is prepared; CSR reports cannot be analyzed directly because they are in PDF format. They are therefore converted into HTML files. Second, a CSR standard is constructed using a topic model. Finally, the CSR rankings are calculated by using a random forest to calculate the importance of features related to CSR activities.

3.2 Data Preparation

This section describes how we extracted textual information from the CSR reports.

HTML Conversion. We used PDFMiner to convert 167 companies' CSR reports into HTML files. PDFMiner is a tool for extracting information from PDF documents. Unlike other such tools, it focuses on obtaining and analyzing textual data. It allows the exact location of a section of text within a page to be obtained, as well as other information such as the fonts used or the number of lines. It includes a PDF converter that can transform PDF files into other text-based formats, such as HTML.

The CSR reports were obtained in the form of PDF files, preventing the text from being extracted directly. In contrast, HTML files enable text to be easily extracted and divided based on tags. We therefore used PDFMiner to convert the PDF files into HTML format.

3.3 Construction of a CSR Standard

Latent Dirichlet Allocation. Latent Dirichlet Allocation (LDA) [5], proposed by Blei in 2003, is a generative probabilistic model, used for collections of discrete data. It analyzes the topics appearing in the documents, using topic and word distributions. Topic distributions express the rates with which particular topics appear in the documents, while the topics themselves are expressed as word distributions, which describe the rates with which words related to that topic appear. LDA estimates both the topic and word distributions.

Applying LDA to the CSR Reports. We created a CSR standard by applying LDA to the CSR reports. Here, we regarded the word distributions created by LDA as a CSR standard. For example, if a topic had words such as "CO_2", "emit", or "factory" in its word distribution, that topic was presumably be related to the environment. LDA enabled us to create a CSR standard based on the CSR reports by determining the topics of the reports.

3.4 CSR Evaluation

After the companies' CSR activities had been quantified using the CSR standard, a random forest was used to rank their contributions to corporate value.

Feature Determination. Topic models learn to estimate the documents' topics by creating suitable features. Each topic is constructed based on a set of relevant words. These features can then be used to determine the topics of documents and hence quantify the CSR activities they describe.

First, the CSR reports were divided into paragraphs, and the nouns were extracted. The topic model then converted each paragraph in each CSR report into a topic distribution, i.e., a vector with dimension equal to the number of topics. The averages of these vectors were then used as features. In this study, we considered three different numbers of topics (10, 30, and 50).

Label. We used labels representing corporate values, which were calculated using the Ohlson model [6].

Ohlson proposed this model, which was based on the discounted dividends model, in 1995. It is used to generate company valuations and uses the income approach. The income approach is a valuation method commonly used for real estate appraisals, and is calculated by dividing the capitalization rate by the net operating income from rental payments. The dividend discount model is based on the idea that a company's stock is worth the sum of all its future dividend payments, discounted back to their present value. The Ohlson model is calculated as follows:

$$V_t = b_0 + \sum_{i=1}^{\infty} \frac{E_t[X^a_{t\,t+i}]}{(1+r)^i} \tag{1}$$

where V_t is corporate value, b_0 is shareholder's equity at book value at the start of the term, $X^a_{t\,t+i}$ is residual profit expected at the time $t + i$, and r is capital cost.

The Ohlson model implies that the economic value shareholders can expect can be calculated by adding the discounted cash flow from future residual profits to the equity at book value. We used this in our study for two reasons.

1. The income approach is appropriate for our CSR analysis, because our aim is to evaluate the effect of CSR over the long term.
2. This model is strongly related to the companies' valuations. Although it can be easily calculated from the accounting information given, it is useful in that it is robust against differences in accounting policy.

One issue with this approach, however, is that we could only obtain limited data about the residual profits. We therefore added a terminal value to the model, which was calculated based on the assumption of zero corporate growth. Although this assumption is not realistic, this should not cause a problem because this model is less reliant on the terminal value than other models are. We calculated the corporate values based on this assumption.

The calculated values were then converted into labels. A random forest was then used to predict these labels, i.e., the CSR activities were assessed in terms of their impact on corporate values. The top 40% of the companies were considered to have positive values, and those of the bottom 40% were negative. The middle 20% was of little use in the analysis, because it would only take a small change to switch from positive to negative, or vice versa.

CSR Ranking. A random forest [7] was used to rank the CSR activities in decreasing order of their contributions to corporate value. Random forests are an ensemble method and use the following algorithm.

1. Use the bootstrap method to extract sets of N samples from the dataset.
2. Create a separate decision tree for each set of samples.
3. Make predictions based on the majority votes of the decision trees.

The importance of a given feature indicates its effectiveness for classification. In this study, the importance describes how different CSR activities contribute to the overall corporate value classification. The CSR activities were then ranked based on these importance values.

4   Results and Discussion

4.1   Topic Classification

Tables 1, 2, and 3 show results for the different topic models. Tables 2 and 3 only show subsets of the topics, for reasons of space.

Table 1. LDA results (10 topics)

Topic 1		Topic 2		Topic 3		Topic 4		Topic 5	
Word	Value	Word	Value	Word	Value	Word	Value	Word	Value
amount	0.039	environment	0.017	product	0.019	thing	0.019	activity	0.028
emit	0.024	activity	0.013	business	0.017	propulsion	0.012	area	0.016
environment	0.021	report	0.011	technology	0.016	of	0.011	support	0.011
reduction	0.019	of	0.010	development	0.013	for	0.011	participation	0.010
Topic 6		Topic 7		Topic 8		Topic 9		Topic 10	
Word	Value	Word	Value	Word	Value	Word	Value	Word	Value
system	0.016	director	0.023	thing	0.017	safety	0.02	yen	0.035
woman	0.016	company	0.019	customer	0.011	implementation	0.019	million	0.018
people	0.016	board	0.018	for	0.009	factory	0.018	thousand	0.018
training	0.016	audit	0.015	customers	0.008	activity	0.017	hundred	0.016

Table 2. LDA results (30 topics)

Topic 1		Topic 2		Topic 3		Topic 8		Topic 10	
Word	Value	Word	Value	Word	Value	Word	Value	word	Value
board	0.024	quality	0.024	thing	0.018	amount	0.069	system	0.032
management	0.022	activity	0.024	food	0.010	emit	0.042	take	0.021
director	0.022	propulsion	0.020	ingredient	0.010	reduction	0.031	child-raising	0.018
audit	0.021	risk	0.016	chemistry	0.009	CO_2	0.026	care	0.015
Topic 13		Topic 18		Topic 25		Topic 26		Topic 28	
Word	Value	Word	Value	Word	Value	Word	Value	Word	Value
chemistry	0.036	safety	0.026	healthcare	0.032	amount	0.038	hold	0.018
administration	0.024	implementation	0.021	cure	0.025	use	0.032	activity	0.015
material	0.021	compliance	0.010	medicine	0.023	water	0.025	company	0.013
preservation	0.016	activity	0.010	patient	0.021	thousand	0.021	support	0.012

Table 3. LDA results (50 topics)

Topic 1		Topic 3		Topic 6		Topic 9		Topic 14	
Word	Value	Word	Value	Word	Value	Word	Value	Word	Value
company	0.039	safety	0.073	environment	0.030	factory	0.181	property	0.030
shareholder	0.024	disaster	0.025	implementation	0.028	business	0.015	patent	0.025
board	0.017	health	0.020	administration	0.021	environment	0.013	pdf	0.020
investor	0.016	sanitary	0.020	activity	0.020	plant	0.009	report	0.016
Topic 19		Topic 26		Topic 40		Topic 41		Topic 44	
Word	Value	Word	Value	Word	Value	Word	Value	Word	Value
social	0.029	support	0.015	authentication	0.038	drug	0.045	%	0.030
thing	0.028	project	0.014	acquisition	0.038	management	0.031	goal	0.025
firm	0.026	manufacture	0.010	factory	0.023	drug-discovery	0.017	2016	0.020
problem	0.016	thing	0.009	iso14001	0.017	medicine	0.016	reduction	0.016

The different CSR activities were divided into topics. For example, Topic 1 in Table 1 represents activities related to reducing CO_2 emissions, because it includes the words "environment" and "emit." Likewise, Topic 6 in Table 1 represents diversity promotion, because it includes the words "woman," and "system." In contrast, Topic 10 features words related to money; hence, it is irrelevant to CSR activity.

Now, we focus on Tables 2 and 3. As more topics are considered, more specific CSR activities appear. For example, Topic 10 in Table 2 is related to childcare leave, a specific aspect of diversity promotion. Using only 10 topics, yields general activities, such as diversity promotion, but with 30 or 50 topics, more specific activities, such as childcare leave, appear. That said, however, it is not helpful to consider too many topics: with more topics, more unrelated topics appear. It is therefore important to use a moderate number of topics.

4.2 Corporate Value

Figure 2 shows a corporate value histogram for the companies considered. Here, the values are concentrated at around 25,000. We assigned positive labels to companies with values above 285,262, and negative labels to those with values below 128,472.

4.3 Classification Accuracy

Table 4 shows the random forest's performance in terms of classification accuracy.

For 30 topics, the accuracy was 0.78, equal to Maki and Feng's result, meaning that the random forest's predictions were mostly correct. In contrast, the accuracy was much lower for 10 topics. We believe this was because the topics were too general and it was unable to properly assess the level of CSR activity. For 50 topics, the accuracy was also slightly lower than for 30, due to the increased number of unrelated topics.

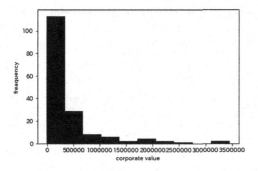

Fig. 2. Corporate value histogram

Table 4. Random forest results

	Accuracy	Precision	Recall
10 topics	0.52	0.53	0.52
30 topics	0.78	0.78	0.78
50 topics	0.74	0.77	0.74

4.4 CSR Ranking

Tables 5 and 6 the top three highest-ranked CSR activities, for Maki and Feng's research (based on ISO 26000) and our method, respectively.

Table 5. CSR activity ranking (Maki and Feng)

Rank	Item	Value
1	Legal compliance office	0.091
2	Biodiversity conservation	0.084
3	Cooperation with NPO and NGO	0.072

Our results differ from those of Maki and Feng. Their highest-ranking topics were legal compliance and biodiversity conservation, but neither of those appears in our results. With 10 topics, Topic 1 (related to the environment) ranked highest, but healthcare ranked highest for 30 and 50 topics. This implies that the companies with the highest corporate values were in the medical sector.

Table 6. CSR activity ranking (current study)

Method	Rank	Topic	Word			Value
LDA (10 topics)	1	1	environment	activity	report	0.176
	2	5	system	woman	people	0.159
	3	4	activity	area	support	0.144
LDA (30 topics)	1	25	healthcare	cure	medicine	0.118
	2	24	reporting	report	2017	0.104
	3	19	human right	effect	and	0.104
LDA (50 topics)	1	43	healthcare	cure	patient	0.083
	2	31	report	organization	society	0.075
	3	32	business	technology	society	0.068

5 Conclusion

In this study, we have used LDA to create a CSR standard. LDA estimates topics based on word distributions, and we used these results to create a CSR standard based on 167 companies' CSR activities. We converted the CSR reports into topic vectors, which were used by a random forest to predict whether particular companies would have high or low corporate values and calculate the importance of different features. The accuracy was highest when we used 30 topics. If the number of topics was too low, the classifier was unable to assess the CSR activities properly, but if it was too high, there were too many unrelated topics. The top-ranked CSR activity was related to healthcare, indicating businesses in the medical sector.

Our research will enable investors to compare companies based on their CSR activities, such as being in the medical sector.

References

1. Orlitzky, M., Schmidt, L.F., Rynes, S.L.: Corporate social and financial performance: a meta-analysis. Org. Stud. **24**(3), 403–441 (2003)
2. Maki, T., Feng, L.: From the ISO- 26000 perspective: does corporate social responsibility influence corporate value? In: Proceedings of the 15th International Conference on Information and Management (2016)
3. CSR liblary.net. http://csr-toshokan.net/. Accessed 15 Apr 2018
4. Nikkei NEEDS. http://www.nikkei.co.jp/needs/. Accessed 15 Apr 2018
5. Blei, D.M., Ng, A.Y., Jordan, M.I.: Latent dirichlet allocation. In: Proceedings of the 15th International Conference on Information and Management, pp. 993–1022 (2013)
6. Ohlson, J.A.: Earnings, book values, and dividends in equity valuation: an empirical perspective. Contemp. Account. Res. **11**(2), 661–687 (1995)
7. Breiman, L.: Random forests. Mach. Learn. **45**(1), 5–32 (2001)

A Weighted Similarity Measure Based on Meta Structure in Heterogeneous Information Networks

Zhaochen Li[1](✉) and Hengliang Wang[2]

[1] College of Mathematics and Physics,
China University of Geosciences, Wuhan 430074, China
zcli@cug.edu.cn
[2] School of Mathematical Sciences, Peking University, Beijing 100871, China
wanghl@pku.edu.cn

Abstract. Evaluating the similarity between two objects in heterogeneous information network is a significant part of information science. The existing meta-structure based similarity measures only consider one meta-structure, which leads to a loss of accuracy. Based on the meta-structure, this paper proposes a weighted method to tackle the problem. We put forward a weighting algorithm that determines the value of weight to each meta-structure according to the set of the user's preferences, and to compute the similarity value, we convert meta-structure into meta-path and use a novel meta-path based similarity measure StruSim. The top-k similarity research experiment is conducted to prove the effectiveness of the novel method. Using the measure nDCG, we conclude that StruSim performs better than PathSim, HeteSim, and AvgSim. And the multiple meta-structure methods are better than BSCSE and unweighted meta-path based methods. At last, we propose an interpolation and derivation method to search the optimal bias factor in StruSim to achieve a better performance.

Keywords: Similarity measure · HIN · Top-k similarity research
Meta-structure

1 Introduction

The studies of heterogeneous information networks (HIN), especially the similarity measure, become more and more popular. Different from homogeneous information networks, HIN includes various objects and links, which is a more accurate abstract of the real world. Making use of the abundant semantics in HIN, the relevance between two objects can be calculated based on the connections in networks, for example, in DBLP [6] dataset, authors are connected by papers they write.

HIN provides a novel model for data mining tasks. In recent years, the developments of this field can be summarized into seven categories [10]: similarity

© Springer Nature Switzerland AG 2018
K. Yoshida and M. Lee (Eds.): PKAW 2018, LNAI 11016, pp. 271–281, 2018.
https://doi.org/10.1007/978-3-319-97289-3_22

search, clustering, classification, ranking, link prediction, recommendation and information fusion. Among these data mining tasks, the similarity measure is the basic task, because it provides a quantitative standard. For example, the recommendation system generates a recommendation list based on the values of similarity measure, and clustering uses similarity measures as the classification criteria. There are multiple meta-path based similarity measures proposed in recent researches, such as PathSim [13], HeteSim [9] and AvgSim [8]. They evaluate the similarity between two objects on the basis of meta-paths of each object. Especially, PathSim focuses on symmetric meta-path, while HeteSim and AvgSim can work out asymmetric meta-path. Weighted PathSim [15] which takes the significance of objects into consideration is proposed for movie recommendation by assigning weight on original meta-path. However, the ability to express information of meta-path is limited. Meta-structure proposed by Huang et al. [2] shows stronger ability to express information than meta-path.

In this paper, we propose a weighted meta-structure framework to evaluate the similarity between two objects in HIN. This framework includes a user-guided weighting algorithm and a method to transform meta-paths to meta-structures. Besides, a novel meta-path-based similarity measure StruSim is put forward. Finally, the top-k similarity ranking experiment is designed to verify the effectiveness of these novel methods.

2   Related Work

Generally, similarity measures are defined based on two structures: meta-path and meta-structure. They contain different amounts and types of information.

For the similarity measures based on meta-path, many scholars have completed comprehensive researches [12]. In 2007, Fouss et al. [1] and others proposed the random-walk (RW) to compute the similarities between nodes of a graph. In 2010, Lao and Cohen [5] introduced a method of combining path-constrained random walks (PCRW) for relational retrieval. In the same years, they proposed a fast query execution method [4] which showed speedups of factors of 2 to 100 with a certain accuracy. In 2011, PathSim [13] was defined by Sun et al. to conduct the top-k similarity search experiment. It focuses on two objects of the same type under a symmetric meta-path. The value of PathSim is determined by the number of meta-path instances between two objects. In 2014, HeteSim [9] was proposed by Shi et al. and it performed well on both the symmetric and asymmetric meta-path. In the same year, Meng et al. [8] put forward AvgSim improved the performance of RW. It performs better than HeteSim in clustering and some other applications. Considering the significance of different objects in meta-path for a given object, weighted PathSim [15] was proposed by Tang et al. This method sets weights on links between two objects on HIN and computes similarity on weighted HIN by using PathSim. However, the meta-path has a disadvantage of less information. In order to solve problems in more complicated information networks, the meta-structure is proposed. In 2016, Huang et al. [2] proposed the definition of meta-structure and introduced three meta-structure based similarity measures: StructCount, Structure

Constrained Subgraph Expansion (SCSE) and Biased Structure Constrained Subgraph Expansion (BSCSE). StructCount is the number of meta-structures between two objects. SCSE, based on the generation process of meta-structures, is essentially an extension of RW on meta-structures. BSCSE adds a bias factor on the basis of SCSE. It is the factor α that combines SCSE and SructCount (The number of instances of $s \in S$) in a unified framework (If $\alpha = 1$, BSCSE reduces to SCSE. If $\alpha = 0$, BSCSE reduces to StructCount).

3 Problem Statement

Similarity measure is to evaluate the relevance between two objects based on HIN. HIN, as an abstract of the real world, is an information network with complex components.

Definition 1 *HIN* [10]. *HIN is a directed graph $G = (V, E)$ with an object type mapping function $\varphi : V \to A$ and a link type mapping function $\phi : E \to R$ in which the types of objects $|A| > 1$ or the types of relations $|R| > 1$.*

The HIN schema [10] $T_G = (A, R)$, defined by object types A and edge types R, is a meta template for a directed graph $G = (V, E)$ shown above. Figure 1(a) shows an example of a bibliography network schema. It expresses all possible link types between objects types.

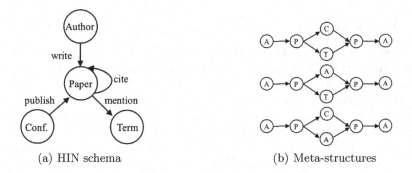

(a) HIN schema   (b) Meta-structures

Fig. 1. HIN schema and meta-structures

Meta-path [10] is defined on a schema $T_G = (A, R)$. It can be denoted by object types as $P = (A_1, A_2, \cdots, A_n)$ and the path instance of P is $p = (a_1, a_2, \cdots, a_n)$ where we use lower-case letters to denote object instances to object types. For example, the physical meaning of the meta-path $APCPA$ is the relationship between two authors who write papers that are published at the same conference, $APTPA$ has the meaning of the relevance two authors who write papers with the same terms, and $APAPA$ means the similarity between two authors who write papers that have the same author.

Definition 2 *Meta-Structure* [10]. *A meta-structure S is a directed acyclic graph, with a single source object v_s and a single target object v_t, defined on a schema $T_G = (A, R)$. It can be denoted as $S = (V, E, v_s, v_t)$, where V is a set of objects and E is a set of edges.*

Figure 1(b) shows three meta-structures that derived from meta-paths shown above. They add more restrictions to papers than corresponding meta-paths.

4   The User-Guided Similarity Algorithm

In the top-k similarity ranking experiment, the selection of meta-structure determines the result. Different users may prefer different ranking results derived from different meta-structures. In order to get the ranking result to meet the requirement of users as much as possible, a general user-guided similarity ranking method need to be proposed.

The process of evaluating the similarity between two objects o_s and o_t can be divided into three steps:

Step 1: Determine the set of meta-structure candidates and calculate the ranking result under each meta-structure candidate
Step 2: Get the weights according to the user's preference
Step 3: Combine the value lists linearly with the weight, and obtain the final weighted ranking result.

4.1   Transformation

This section introduces a transform method to convert a meta-structure into a meta-path, so that meta-structure based similarity measures can be calculated by meta-path based similarity measures.

Note that the set of meta-structure candidates is $MS = \{S_1, S_2, \cdots, S_m\}$, where S_i is the i-th meta-structure candidate. Based on the concepts of meta-structure layers, we denote that the layers of S_i are $L_i^1, L_i^2, \cdots, L_i^{total-layer}$ where *total-layer* is the number of layers of S_i. The transform method is combining the nodes of each layer into one combinatorial node and the number of instances of each combinatorial node is the product of the original nodes' scales. For example, the converted meta-paths of Fig. 1(b) are $P_1 : A \to P \to (C, T) \to P \to A$, $P_2 : A \to p \to (A, T) \to P \to A$, $P_3 : A \to P \to (C, T) \to p \to A$, and the instance sets of combinatorial nodes are $\{(c, t) | c \in C, t \in T\}$, $\{(a, t) | a \in A, t \in T\}$, $\{(c, a) | c \in C, a \in A\}$, respectively.

4.2   StruSim

After transforming meta-structures to meta-paths, the similarity can be obtained by using relevance measures based on meta-path. We propose a novel similarity measure based on meta-path.

Definition 3 *StruSim. Given an HIN $G = (V, E)$, a meta-path $P = (A_1, A_2, \cdots, A_{n+1})$, a source object o_s and a target object o_t. Denote that A_i is the i-th layer of P, o_i represents the object at the i-th layer, StruSim of o_i is defined as follows:*

$$StruSim(o_i, o_t) = \frac{\sum_{o'_{i+1} \in \{o_{i+1}|(o_i, o_{i+1}) \in E\}} StruSim(o'_{i+1}, o_t|P)}{|\{o_{i+1}|(o_i, o_{i+1}) \in E\}|^\alpha} \quad (1)$$

where $\{o_{i+1}|(o_i, o_{i+1}) \in E\}$ is a set of A_{i+1}'s instances that have direct links with o_i, α is a bias factor ranging from 0 to 1. Especially, o_1 is o_s. The terminating condition of the recursive operation is $i = n + 1$. $StruSim(o_i, o_t|P) = 1$ if and only if $o_i = o_t$, otherwise $StruSim(o_{n+1}, o_t|P) = 0$.

Figure 2(a) is a toy HIN and Fig. 2(b) shows an expanding tree under the meta-path $AP(C, T)PA$.

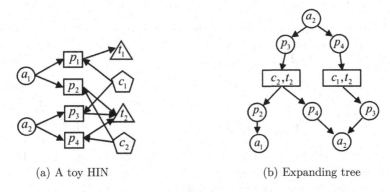

(a) A toy HIN   (b) Expanding tree

Fig. 2. Example

In Fig. 2(b), $StruSim(a_2, a_1) = \frac{(1/2)^\alpha + 0}{2^\alpha}$ and $StruSim(a_1, a_1|P) = \frac{(1/2)^\alpha + 1}{2^\alpha}$. If $\alpha = 1$, the result of StruSim is the same as Random Walk (RW). If $\alpha = 0$, StruSim reduces to the number of path instances between two objects (i.e. PathCount). Therefore, StruSim is a unified framework of RW and PathCount.

Actually, the similarity between two identical objects should be the largest. The ideal state of the model in Fig. 2(b) is that there is a significant difference between $StruSim(a_2, a_1|P)$ and $StruSim(a_1, a_1|P)$, so the bias factor should be adjusted to distinguish them. In Fig. 2(b), the difference between the value of $StruSim(a_2, a_1|P_1)$ and $StruSim(a_1, a_1|P_1)$ is $1/2^\alpha$ which has a ratio of 2^α to $StruSim(a_2, a_1|P_1)$ and $\frac{2^\alpha}{1+2^\alpha}$ to $StruSim(a_1, a_1|P_1)$. In order to achieve the ideal state, 2^α should be larger while $\frac{2^\alpha}{1+2^\alpha}$ should be smaller (i.e. α can't be too large or too small). Therefore, there exists the most suitable α for StruSim.

4.3 Weighting Algorithm

After obtaining the ranking lists $\{List_1, List_2, \cdots, List_m\}$ w.r.t. meta-structure candidates, the weight of each ranking list should be set. Actually, users can select any suitable weighting algorithms in this part such as the algorithm mentioned in [14]. In this paper, we provide a novel simple weighting method.

We denote by $U = \{U_1, U_2, \cdots, U_r\}$ as a set of some particular objects to store the user's preferences, and the user-guided weight can be calculated by a weighting algorithm.

It first initializes the weight set $\mathbf{w}$ and then uses the scoring system to get the weight values. For each preference U_i, there is a ranking value for each ranking list $List_j$. We sort the ranking value in an ascending order, and denote by $rank_{ij}$ as the ranking of $List_j$ w.r.t. U_i. The scoring method is determined by the ranking value in two conditions. The first condition is that there aren't any juxtapositions in the ranking lists. In this case, the score vector of U_i is

$$\mathbf{score}_i = (m + 1 - rank_{i1}, m + 1 - rank_{i2}, \cdots, m + 1 - rank_{im}) \qquad (2)$$

while if there exist juxtapositions in the ranking lists, the scores of the lists that have the same ranking value will be the average of rankings. In general, we suppose that $List_s, List_{s+1}, \cdots, List_{s+k}$ have the same ranking value, the score vector of U_i is

$$\mathbf{score}_i = \begin{cases} m + 1 - (rank_{is} + rank_{is+k})/2, & j \in [s, s+k] \\ m + 1 - rank_{ij}, & \text{otherwise} \end{cases} \qquad (3)$$

Finally, we can get the weight vector $\mathbf{w} = \sum_{i=1}^{r} \mathbf{score}_i$, and it will always be normalized in practical application.

For example, there are two ranking lists of authors: $List_1 : a_1, a_2, a_4, a_3$ and $List_2 : a_2, a_1, a_3, a_4$, and $U = \{a_2, a_4\}$. For the user's preference a_2, the ranking value is 2 in $List_1$ and 1 in $List_2$, and then we sort them and get $\mathbf{score}_1 = (1, 2)$. For the user's reference a_4, the score vector is $\mathbf{score}_2 = (2, 1)$. The normalized weight vector is $\mathbf{w} = (0.5, 0.5)$.

5 Experiments

In order to examine the effectiveness of the novel similarity measure defined above, we design a top-k similarity search experiment [8].

The experiment can be divided into two parts. The first part is comparing the effectiveness among PathSim, HeteSim, AvgSim and StruSim after converting the meta-structure based similarity into meta-path based similarity. The second part is consist of two tests: (1) Compare the multiple meta-structures research and the that the single meta-structure research, (2) Compare the result of the multiple meta-structures research and that of the weighted meta-path research. At the end of the experiment, we analyze the results of ranking lists by using the measure nDCG [3] (Normalized Discounted Cumulative Gain).

5.1   Database

We use the DBLP [7] dataset and DouBan movie (DBM) [11] dataset as the test data. DBLP is a computer science bibliography website for open bibliographic information. The dataset we use contains four types of objects: 14475 authors, 20 conferences, 14376 papers and 8920 terms. The HIN schema we use is the same as Fig. 1(a) shows, except for the citation link between two papers because of the lack of information. DBM consists of 6971 movies, 1000 users, 789 directors, 1000 actors and 36 movie types.

5.2   Effectiveness

We set the query as John M. Prager whose ID number in the dataset is 4139, and set the user's preferences are Pablo A. D. and Krzysztof Czuba whose ID number are 4005 and 4140 respectively. The main purpose of the experiment is to find the top-10 most similar authors to John M. Prager.

Comparing StruSim with Other Meta-Path-Based Measures. When a meta-structure is converted into a meta-path, other measures such as PathSim, HeteSim and AvgSim can be applied on the same meta-path, and after getting the ranking lists, the rest of procedure are also the same.

A case study is shown in Table 1, which is applied on DBLP dataset, under three meta-structures shown in Fig. 1(b), and the weights of meta-structures are shown in Table 2. In Table 1, the first place of the most similar author in the three ranking lists is John M. Prager himself, which is consistent with common sense. HeteSim and StruSim show the same ranking result, so StruSim has the similar efficiency with HeteSim in some degree.

Table 1. Ranking lists

Rank	PathSim	HeteSim	AvgSim	StruSim($\alpha = 0.3$)
1	John M. Prager	John M. Prager	John M. Prager	John M. Prager
2	Jennifer Chu-Carroll	Jennifer Chu-Carroll	Jennifer Chu-Carroll	Jennifer Chu-Carroll
3	Dragomir R. Radev	Dragomir R. Radev	Dragomir R. Radev	Dragomir R. Radev
4	Zhu Zhang	Pablo A. D.	Pablo A. D.	Pablo A. D.
5	Sasha B. Goldensohn	Krzysztof Czuba	Krzysztof Czuba	Krzysztof Czuba
6	Hong Qi	David A. Ferrucci	David A. Ferrucci	David A. Ferrucci
7	Weiguo Fan	Zhu Zhang	Sarah Luger	Zhu Zhang
8	Zhiping Zheng	Sasha B. Goldensohn	Zhu Zhang	Sasha B. Goldensohn
9	Pablo A. D.	Hong Qi	Sasha B. Goldensohn	Hong Qi
10	Krzysztof Czuba	Weiguo Fan	Hong Qi	Weiguo Fan

To evaluate the rationality of the three ranking lists, we label each ranking object with relevance score as three levels: 1-low-relevant, 2-some-relevant and

Table 2. Weights of meta-structures

Method	AP(CT)PA	AP(AT)PA	AP(CA)PA
PathSim	0.3	0.3	0.3
HeteSim	0.5	0.25	0.25
AvgSim	0.5	0.25	0.25
StruSim	0.5	0.25	0.25

Table 3. nDCG of each ranking list on DBLP

Measure	PathSim	HeteSim	AvgSim	StruSim
nDCG	0.8906	0.9938	0.9885	0.9938

3-high-relevant. Then we use nDCG to assess the quality of ranking lists by comparing their results with relevance scores, and the result is shown in Table 3. As we can see, the ranking quality by using StruSim and HeteSim are the best among the four methods, which proves the efficiency of StruSim.

As for DBM dataset, we input the 204th director and set the preferences as the 420th director and the 663th director. After computing the ranking lists of directors, we get nDCG in Table 4. This result shows that StruSim has the highest accuracy.

Comparing the Novel Method with BSCSE. To evaluate the effectiveness of the weighting algorithm, we set a case study to compare it with the single meta-structure test. Generally, we use BSCSE under the meta-structure AP(CT)PA and obtain the ranking list.

After that, we use nDCG to evaluate the ranking quality of BSCSE and user-guided multiple meta-structures based measure, the result is 0.9881, which is smaller than that of the novel method. Therefore the method that considers multiple meta-structures contributes to generating a higher quality ranking list.

Comparing the Novel Method with Unweighted transformation Method. The first experiment shows that HeteSim and StruSim have the same ranking results. To further evaluate the performance of these two algorithms, this section designs a case study to compute the results of unweighted conversion method. After converting meta-structures to meta-paths, we use HeteSim and StruSim to get ranking lists without using weighting algorithm.

Table 4. nDCG of each ranking list on DBM

Measure	PathSim	HeteSim	AvgSim	StruSim
nDCG	0.7560	0.8174	0.8833	0.9495

The nDCG of HeteSim and StruSim are respectively 0.8906 and 0.9938. Comparing the values in Table 3, we conclude that HeteSim performs better with the help of the weighting algorithm and the conversion method based on StruSim has a stronger stability than that of HeteSim.

The Effect of the Bias Factor α. We have proved that there exists the most suitable bias factor α for StruSim. We set the step length to 0.1, and calculate the nDCG of ranking lists under different bias factor in $[0, 1]$ interval (Table 5).

Table 5. Bias factors and the corresponding nDCG

α	0	0.1	0.2	0.3	0.4	0.5	0.6	0.7	0.8	0.9	1.0
nDCG	0.8906	0.9938	0.9938	0.9938	0.9885	0.9885	0.9684	0.9684	0.9684	0.9189	0.9189

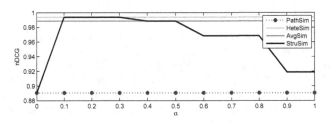

Fig. 3. Varying α

To reflect the effect of α on the ranking results more intuitively, we get Fig. 3. We can see that HeteSim, StruSim and AvgSim all have an absolute advantage over PathSim, and the performance of StruSim is better than AvgSim when $\alpha \in [0.1, 0.4]$. In addition, the trend of StruSim's curve increases first and then decreases, which can be regarded as a single peak parabola. Based on this assumption, we can search the optimal value of α by interpolation and derivation.

Interpolation and Derivation Method. Based on the data in Table 5, we use three spline interpolation to get the relation of α and nDCG. According to the curve of interpolating function, we know that the range of optimal α can be reduced to $[0.1, 0.2]$, so we can focus on the function in this interval:

$$f(x) = 22.4173(x - 0.1)^3 - 5.16(x - 0.1)^2 + 0.2918(x - 0.1) + 0.9938$$

We get the maximum point of $f(x)$ in $[0.1, 0.2]$ by derivation as $\alpha = 0.1374$, and then we set $\alpha = 0.1374$ and get the corresponding ranking result is the same as that of $\alpha = 0.3$. Therefore, we have found out the best result under the novel method.

6   Conclusions and Future Work

In this paper, we introduce a novel method to evaluate the similarity between two objects in HIN. This method takes multiple meta-structures into consideration and weights them by the user's preferences. To calculate the value of similarity under each meta-structure, we convert meta-structure into meta-path and propose a novel meta-path based similarity measure StruSim. The top-k similarity ranking research experiment on DBLP dataset demonstrates the effectiveness of the novel method and we, at the end of the experiment, propose a method to find out the optimal bias factor defined in StruSim to obtain the best ranking result.

In the future, we will study the algorithm for automatically generating meta-structures from knowledge base and the optimization method to speed up the recursion in StruSim. We will also examine the effectiveness of the novel method in different applications, such as clustering and recommendation.

References

1. Fouss, F., Pirotte, A., Renders, J.M., Saerens, M.: Random-walk computation of similarities between nodes of a graph with application to collaborative recommendation. IEEE Trans. Knowl. Data Eng. **19**(3), 355–369 (2007)
2. Huang, Z., Zheng, Y., Cheng, R., Sun, Y., Mamoulis, N., Li, X.: Meta structure: computing relevance in large heterogeneous information networks. In: Proceedings of the 22nd ACM SIGKDD International Conference on Knowledge Discovery and Data Mining, pp. 1595–1604. ACM (2016)
3. Järvelin, K., Kekäläinen, J.: Cumulated gain-based evaluation of IR techniques. ACM Trans. Inf. Syst. (TOIS) **20**(4), 422–446 (2002)
4. Lao, N., Cohen, W.W.: Fast query execution for retrieval models based on path-constrained random walks. In: Proceedings of the 16th ACM SIGKDD International Conference on Knowledge Discovery and Data Mining, pp. 881–888. ACM (2010)
5. Lao, N., Cohen, W.W.: Relational retrieval using a combination of path-constrained random walks. Mach. Learn. **81**(1), 53–67 (2010)
6. Ley, M.: DBLP computer science bibliography (2005)
7. Meng, C., Cheng, R., Maniu, S., Senellart, P., Zhang, W.: Discovering meta-paths in large heterogeneous information networks. In: Proceedings of the 24th International Conference on World Wide Web, pp. 754–764. International World Wide Web Conferences Steering Committee (2015)
8. Meng, X., Shi, C., Li, Y., Zhang, L., Wu, B.: Relevance measure in large-scale heterogeneous networks. In: Chen, L., Jia, Y., Sellis, T., Liu, G. (eds.) APWeb 2014. LNCS, vol. 8709, pp. 636–643. Springer, Cham (2014). https://doi.org/10.1007/978-3-319-11116-2_61
9. Shi, C., Kong, X., Huang, Y., Yu, P.S., Wu, B.: HeteSim: a general framework for relevance measure in heterogeneous networks. IEEE Trans. Knowl. Data Eng. **26**(10), 2479–2492 (2014)
10. Shi, C., Yu, P.S.: Heterogeneous Information Network Analysis and Applications. Data Analytics. Springer, New York (2017). https://doi.org/10.1007/978-3-319-56212-4

11. Shi, C., Zhang, Z., Luo, P., Yu, P.S., Yue, Y., Wu, B.: Semantic path based personalized recommendation on weighted heterogeneous information networks. In: Proceedings of the 24th ACM International on Conference on Information and Knowledge Management, CIKM 2015, pp. 453–462. ACM (2015)
12. Sun, Y., Han, J.: Meta-path-based search and mining in heterogeneous information networks. Tsinghua Sci. Technol. **18**(4), 329–338 (2013)
13. Sun, Y., Han, J., Yan, X., Yu, P.S., Wu, T.: PathSim: meta path-based top-k similarity search in heterogeneous information networks. PVLDB **4**(11), 992–1003 (2011)
14. Sun, Y., Norick, B., Han, J., Yan, X., Yu, P.S., Yu, X.: Integrating meta-path selection with user-guided object clustering in heterogeneous information networks. In: Yang, Q., Agarwal, D., Pei, J. (eds.) The 18th ACM SIGKDD International Conference on Knowledge Discovery and Data Mining, KDD 2012, Beijing, China, 12–16 August 2012, pp. 1348–1356. ACM (2012)
15. Tang, Z.P., Yang, Y., Bu, Y.: Weighted-pathSim: similarity measure for plot-based movie recommendation

Author Index

Printed in the United States
By Bookmasters